Management, Change, Strategy and Positive Leadership

Series Editors
Satinder Dhiman
School of Business
Woodbury University
Burbank, CA, USA

Joan Marques
School of Business
Woodbury University
Burbank, CA, USA

This series publishes authored and edited books that explore management, change, strategy and positive leadership in the disciplines of business, health care, education and various other allied fields. It presents a holistic thematic schema of management and leadership domains that include human resource development, workplace flourishing, and spiritual leadership, among others. Global in its intent, scope and purpose, the series editors invite contributions from researchers, practitioners and scholars in all continents in the aforementioned fields and disciplines. While primarily a business, management and leadership series, interdisciplinary and multidisciplinary titles are welcome.

Nadia Singh • Mai Chi Vu
Irene Chu • Nicholas Burton
Editors

Faith Traditions and Sustainability

New Views and Practices for Environmental Protection

Springer

Editors
Nadia Singh
Newcastle Business School
Northumbria University
Newcastle upon Tyne, UK

Mai Chi Vu
Newcastle Business School
Northumbria University
Newcastle upon Tyne, UK

Irene Chu
School of Management
Newcastle University
Newcastle upon Tyne, UK

Nicholas Burton
Newcastle Business School
Northumbria University
Newcastle upon Tyne, UK

ISSN 2662-3080 ISSN 2662-3099 (electronic)
Management, Change, Strategy and Positive Leadership
ISBN 978-3-031-41244-8 ISBN 978-3-031-41245-5 (eBook)
https://doi.org/10.1007/978-3-031-41245-5

This Springer imprint is published by the registered company Springer Nature Switzerland AG
The registered company address is: Gewerbestrasse 11, 6330 Cham, Switzerland

Paper in this product is recyclable.

Preface

We are facing unprecedented environmental challenges in the form of heightened climate change, increasing levels of water and air pollution and biodiversity losses. The United National Climate Change Conference in 2021 highlighted the urgency to bring about transformative changes in our responses to environmental degradation by both national governments and corporations to create sustainable economies, which are resilient to environmental challenges. To support the United Nations Sustainable Development Goals within the 2030 Agenda, the United Nations Environment Program launched the Faith for Earth Initiative in 2017, which facilitates the engagement with faith-based organizations to collectively achieve the Sustainable Development Goals. This shows faith traditions and faith-based organizations play a significant role at the global, regional, and local levels in addressing issues of climate change, pollution, and biodiversity (Ambassador Mussie Hailu, Director of Global Partnership, United Nations Initiative).

Faith traditions shape human behaviour, affect social cohesion, consumption, mitigation, and adaptation initiatives, and remain crucial to address environmental issues. This book is relevant and timely as it explores various environmentally sustainable initiatives of organizations rooted in different faith traditions and practices. The contributions in this volume introduce new approaches to environmental sustainability, which depart from mainstream sustainability frameworks. The chapters in this book include the following faith traditions that have been quoted in the 45th World Environment Day in 2019 (United Nations Environment Programme) as influential faith traditions that speak to the interconnectedness between faith and the environment.

> Baha'i: "Nature is God's Will and is its expression in and through the contingent world." (Tablets of Bahá'u'lláh, p. 142)

> Buddhism: "Our ancestors viewed the earth as rich and bountiful, which it is. Many people in the past also saw nature as inexhaustibly sustainable, which we now know is the case only if we care for it." (Dalai Lama, 1990a)

> Christianity: "We must treat nature with the same awe and wonder that we reserve for human beings. And we do not need this insight in order to believe in God or to prove his

existence. We need it to breathe; we need it for us simply to be." (Ecumenical Patriarch Bartholomew, 2010)

Confucianism: "… sustainable harmonious relationship between the human species and nature is not merely an abstract ideal, but a concrete guide for practical living." (International Confucian Ecological Alliance, 2015)

Judaism: "And God said: Behold, I have given you every herb yielding seed, which is upon the face of all the earth, and every tree, in which is the fruit of a tree yielding seed—to you it shall be for food." (Gen 1:29)

Sikhism: "You, yourself created the Universe, and You are pleased… You, Yourself the bumblebee, flower, fruit and the tree." (Guru Granth Sahib, Maru Sohele, page 1020)

Newcastle upon Tyne, UK Nadia Singh
Newcastle upon Tyne, UK Mai Chi Vu
Newcastle upon Tyne, UK Irene Chu
Newcastle upon Tyne, UK Nicholas Burton

Acknowledgments

We are grateful to Prof. Ron Beadle at Northumbria University for inspiring us to work on this book. Ron has been a constant source of inspiration and support through this long and arduous journey. We would also like to thank our publisher, Springer Nature, for endowing us with this valuable opportunity.

The idea of this work first originated in the "Faith in Business" conference organized by Ron at Northumbria University in 2021 wherein we were able to explore common strands of research and foster fruitful collaborations with scholars and practitioners from different faith traditions. This volume represents the voices of many different scholars from across different regions and traditions.

Contents

About the Editors

Nadia Singh is an Assistant Professor of Economics at Northumbria University, UK, and is a Fellow of the Higher Education Academy of the UK. Nadia completed her PhD in Economics at Oxford Brookes Business School (Oxford, UK). She has published her work in journals like the *World Development*, *Journal of Development Studies* and *Gender*, *Work and Organisation*. Her research interests are centered on ecological and gender economics, as well as the influences of Sikh philosophy on issues of environmental governance and organizational sustainability.

Mai Chi Vu is an Assistant Professor of Responsible Business at Newcastle Business School, Northumbria University, UK. She holds a PhD in Management from Durham University Business School (Durham, UK). She publishes regularly in the *Journal of Business Ethics*, *Management Learning*, *Journal of Management Inquiry*, *European Management Review*, *Journal of Business Research*, and the *International Journal of Human Resource Management*. Her research interests focus on organizational mindfulness, business ethics, meaningful work, and workplace spirituality.

Irene Chu is a Lecturer of International Management at Newcastle Business School, Newcastle University, UK. She received her PhD from Durham University Business School. She has publications in several journals including the *Journal of Business Ethics*, *British Journal of Management*, *Technological Forecasting and Social Change*, and *Research in International Business and Finance*. Her research interests include business ethics, especially virtue ethics, cross-cultural and interdisciplinary studies, and institutional logics.

Nicholas Burton is an Associate Professor of Entrepreneurship, Innovation and Strategy, at Northumbria University, UK. He has published in the *Journal of Business Ethics*, *Management Learning*, *Journal of Management History*, *Review of Managerial Science*, and *Strategic Change*.

Contributors

Nicholas Burton Newcastle Business School, Northumbria University, Newcastle upon Tyne, UK

Susan S. Case Case Western Reserve University, Cleveland, OH, USA

Irene Chu School of Management, Newcastle University, Newcastle upon Tyne, UK

Allan Discua Cruz Centre for Family Business, Lancaster University Management School, Lancaster, UK

George Gotsis National and Kapodistrian University of Athens, Athens, Greece

Sheila K. Hanson University of North Dakota, Grand Forks, ND, USA

Viet-Ngu Hoang QUT School of Business, Brisbane, Australia

Florence Ingram TKSL School of Buddhist Science, Brisbane, QLD, Australia

M. Isabella Cavalcanti Junqueira LeTourneu University, Longview, TX, USA
Montreat College, Montreat, NC, USA

Ksenia Keplinger Max Planck Institute for Intelligent Systems, Stuttgart, Germany

Legha Momtazian The University of Gloucestershire, Cheltenham, UK

Nadia Singh Newcastle Business School, Northumbria University, Newcastle upon Tyne, UK

Tsewang Topden Sera Jey Monastery University, Bylakuppe, India

Mai Chi Vu Newcastle Business School, Northumbria University, Newcastle upon Tyne, UK

Abbreviations

ARC	Alliance for Religion and Conservation
CBA	Cost-Benefit Analysis
GHG	Greenhouse Gases
EPGN	European Green Pilgrimage Network
GPN	Global Pilgrimage Network
LK&C	Loving, Kindness and Compassion
RSB	Rule of St Benedict
OSB	Order of St Benedict
SDGs	Sustainable Development Programme
SGGS	Sri Guru Granth Sahib
UN	United Nations
UNDP	United Nations Development Programme
UNEP	United Nations Environment Programme
UNFCC	United Nations Framework on Climate Change
WCED	World Commission on Environment and Development
WEF	World Economic Forum
WWF	World Wildlife Fund

Part I
Introduction

Chapter 1
Faith Traditions and Sustainability: Key Discourses and Emerging Field

Nadia Singh, Mai Chi Vu, Irene Chu, and Nicholas Burton

Abstract Historically, academic debates on the sustainable development discourse have tended to sideline the influence of religion faith traditions. However, in the last two decades, with the growing instability in the social, economic, and environmental realms of the global economy, there has been intensive search for new paradigms and frameworks to guide human understanding of the interrelationship between human well-being and environmental protection and foster new forms of environmental activism at the ground level. This work intends to contribute to this emerging area of research by presenting how faith traditions can help to provide alternative explanations to deal with contemporary environmental challenges by creating alternative models of production and consumption in society, as well as providing individual motivation for pro-environmental behaviors. The chapters presented in this edited volume weave together the economic, ethical, cultural, and societal dimensions of varied Eastern and Western faith traditions and discusses their applicability to contemporary environmental problems. The work presents three main pathways through which faith-based traditions can help in steering mainstream sustainable development discourses in a new direction – promotion of ethical values, fostering new forms of ecological activism, and inculcating pro-environmental behaviors. The introductory chapter presents an analytical basis for understanding the causes behind sustainability challenges from a faith-based perspective and how these would lead to alternative adaptation and mitigation policies, discussed in the subsequent chapters.

N. Singh (✉) · M. C. Vu · N. Burton
Newcastle Business School, Northumbria University, Newcastle upon Tyne, UK
e-mail: nadia.singh@northumbria.ac.uk; mai.c.vu@northumbria.ac.uk; n.burton@northumbria.ac.uk

I. Chu
School of Management, Newcastle University, Newcastle upon Tyne, UK
e-mail: Irene.Chu@newcastle.ac.uk

N. Singh et al. (eds.), *Faith Traditions and Sustainability*, Management, Change, Strategy and Positive Leadership, https://doi.org/10.1007/978-3-031-41245-5_1

Keywords Faith traditions · Religion · Environmental pressures · Ecological activism · Sustainable development · Pro-environmental attitudes

1.1 Introduction

As ecological problems in society continue unabated both in terms of their magnitude and complexity, several scholars have pressed for the need to develop new paradigms and theoretical approaches (Vu et al., 2022; Singh, 2022 b; Singh & Singh, 2019; Johnston, 2014b; Jenkins & Chapple, 2011; Gottlieb, 2010) to guide human understanding of the environmental question and provide motivation for pro-environmental action. The severity of environmental problems is best illustrated by the case of greenhouse gas emissions in the environment, which are leading to an increasing threat of climate change and global warming. It is feared that climate change may manifest into devastating consequences for the natural environment through catastrophic events such as increasing acidity of seas and threat to arctic wildlife. According to the World Wildlife Fund (2022), 97 percent of the scientist community believe that the most severe repercussions of climate change are rooted in human action such as burning of fossil fuels and massive deforestation. Given the increasing gravity of this problem, questions are being raised on the efficacy and suitability of mainstream environmental approaches in resolving the ecological crisis. There are also increasing calls to develop alternative paradigms and approaches to steer pro-environmental attitudes and behaviors (Singh, 2022a, b).

It is being increasingly propounded that existing technocratic and market-based approaches, rooted in mainstream economic theory, are "insufficient" and "piecemeal" in nature (Singh, 2022a; Ehrenfeld & Hoffman, 2013; Daniels, 2010) and do not adequately address the gravity of the environmental crisis. This is attributed to the fact that mainstream environmental economy theories essentially draw on the theory of "externalities" (Baumol and Oates 1988). Within this framework environmental degradation is regarded as an external cost in the process of economic growth and industrialization, which is not reflected in the market price of goods and services (Singh, 2022a). These approaches contend that the environmental crisis may be resolved through creation of "green markets," employing a range of economic measures such as carbon taxes, government subsidies, direct state regulation, and tradeable pollution permits (Singh, 2022a). However, these approaches are being critiqued as a "reductionist approach" because they do not recognize "the socio-economic and cultural parameters in modern capitalist societies" (Singh, 2022a, p. 324) which have led to these unsustainable patterns of economic development. These approaches however fail to consider humanity's dialectical connection with the natural environment and the interconnectedness between environmental well-being and larger societal well-being. As a result, these policies have failed to demonstrate tangible benefits for the natural environment, especially so in developing country contexts, characterized by weak governance and regulation measures (Abdelzaher & Abdelzaher, 2017). In this context, the World Economic Forum (WEF) advocated, "a new economic system that combines 'productivity,' 'people' and 'planet' targets will provide a new compass for the future direction of economic growth" (Schwab & Zahidi, 2020, p.).

Sustainability remains a "fuzzy" concept (Markusen, 2003, p. 702) with multiple interpretations, conceptualizations, and applications, rooted in different schools of thought. Several scholars have pointed out that the term sustainability lacks a precise definition and is open to several contesting interpretations, which makes it difficult to implement this concept in ground-level imperatives (Narayanan, 2013; Jenkins & Chapple, 2011; Johnston et al., 2007). Sustainability is alternatively considered to be an "ethic" (Johnston, 2014a), "a religion" or a "sacred cow" (Summer, 2005, p. 77), and a "development paradigm" (Briassoulis, 1999, p. 899). In its simplest form, sustainability is conceptualized as "the ability to exist and flourish for lengthy time frames" (Starik & Rands, 1995, p. 910).

On one hand, neo-classical environmental economics subscribe to the concept of "weak sustainability" and contend that natural capital may be replaceable with human-made and manufactured capital in the long run through sufficient technological progress. On the other hand, greens and ecological economists remain firmly rooted in a "strong sustainability" doctrine and argue that natural capital has an intrinsic value of its own and is not substitutable with manufactured/human-made capital (Singh & Singh, 2019). These multiple definitions of sustainability reflect on the stark differences in conceptualization and treatment of the interconnections and interrelationships between the environmental, economic, and social realms of life (Landrum & Ohsowski, 2018). However, it is becoming increasing apparent that the realization of sustainability or long-term survival requires planned strategy, which will only be fructified through the integration of two domains: (a) modification in individual consciousness and attitudes toward the natural environment and (b) systemic changes in assumptions that society holds in maximizing utility or material satisfaction.

The term sustainable development first gained prominence in the year 1987 with the publication of the Brundtland Report by the World Commission on Environment and Development (1987). The Brundtland Report defined sustainable development as development that "meets the needs of the present without compromising the ability of future generations to meet their own needs" (WCED, 1987, p. 8). On paper, this definition seeks to integrate the economic, social, and ecological realms of life into a single, unifying concept. However, over time this conceptualization of sustainability is being critiqued as "ambiguous" (Johnston et al., 2007), "too loose" (Hallin et al., 2021), lacking a precise conceptualization of essential needs (Banerjee & Bonnefous, 2011), and failing to identify the precise policy mechanisms needed to transition to long-run sustainability. In this regard, Johnston et al. (2007, p. 60) remarks, "The term sustainability has become an oxymoron … a vehicle to perpetuate the many corporate interests, while giving the impression of adherence to environmentally sound principles."

In recent years, it is being argued that existing sustainability doctrines are "reductionist in nature" as they have lowered the concept to an "accounting or a cost-benefit exercise" (Intezari, 2015, p. 620) without consideration of the "notions of morality, power structures, institutional logics and cultural factors" that influence the transition to a sustainable development discourse. It is therefore being increasing recognized that a sustainable development pathway must address society's encompassing beliefs and value systems, including spiritual and religious values (Singh, 2022c; Singh & Clark, 2016; Seiple, 2016; Prill, 2015; Marques, 2010; Welford, 1998). As more than 84 percent of the global population (Pew Research Centre, 2017) follows some form of religious/spiritual beliefs, the incorporation of the values and insights from different faith traditions can "enhance prospects for sustainability" (Abeydeera et al., 2016, p. 44). These inclusions can foster a more holistic comprehension of the complex interlinkages between human well-being, environmental protection, and overall economic progress. At the ground level, these theoretical developments could challenge the "business as usual" attitude and encourage business organizations as well as national governments to develop more holistic policies and programs to tackle sustainability issues (Bouckaert & Zsolnai, 2012).

Traditionally, the role of faith influences on sustainable development had been treated with indifference and suspicion. However, in the last few years, existing scholarship in the area has expanded considerably (e.g., Vu et al., 2022; Johnston, 2014b; Ehrenfeld & Hoffman, 2013; Summer, 2005), working across many disciplines such as cultural studies, anthropology, sociology, philosophy, economics, psychology, sociology, and business/management studies. There has also been much debate over issues of methodology, how should religion be conceptualized, and how do religions ideas and doctrines shape relations between the individual and the natural environment (for review see Narayanan, 2013). These interpretative debates signal the development of a new and robust field of research which embeds multiple research programs. A common thematic concern that emerges from these discussions is how religious thoughts and practices can be best re-interpretated and re-employed to influence environmental attitudes and practices at the community level. This work contributes to this nascent field of research by bringing together scholars from various Eastern and Western faith traditions, as well as various disciplines including economics, business ethics, human resources, sociology, and management science to address the multiple ways in which faith traditions can foster new understanding of the sustainability question.

1.2 Faith and Sustainability: Conceptual Moorings

At the outset, the realization of sustainability involves creating enabling condition for achieving acceptable levels of human welfare across space (intragenerational) dimensions and time (intergenerationally) (Daniels, 2007). This necessitates the maintenance of three main forms of capital at constant levels (Daniels, 2007). These three forms of capital are defined as follows by Munn (2002):

1. Manufactured capital comprises of goods and services, which are intended to either provide further inputs into the process of production or be used for consumption to satisfy material wants.
2. Natural capital consists of natural resources such as air, water, soil plants, vegetation, and biodiversity, derived directly from nature.
3. Social capital comprising of social networks, social relations, institutions, and community relations.

Strong social capital is considered a necessary prerequisite for achievement of human well-being (Sandefur & Laumann, 2009) and realization of long-term sustainability imperatives. However, in dominant sustainability frameworks such as "triple bottom line" approach and "three pillars" approach, social capital tends to be relegated to a supporting role, to achieve economic and environmental aspects of sustainable development. Thus, the subjective and sociopsychological aspects of human well-being have not been accorded primacy in mainstream sustainable development paradigm. Some scholars (Singh, 2022a; Singh & Singh, 2019; Daniels, 2010; Grzeda, 2019) attribute this to the fact that sustainable development continues to be embedded in neoclassical economic theory wherein growth and profitability are accorded primacy and environmental/social concerns are only looked after to an extent that they do not become impediments in the path of capital accumulation and economic growth.

However, human well-being is an increasingly complex and subjective concept (Narayanan, 2013) which depends on multiple factors including an individual's mental/psychological cognizance, physical health, as well as their relations with the community, sociocultural environmental, and the natural environment. The philosophical paradigms espoused by various faith traditions can help to unpack the fundamental contestations between human welfare, environmental protection, and economic progress and help to charter an inclusive sustainable development pathway in society. In this context, scholars like Rakodi (2012, p. 625) referred to religion as the "missing ingredient" in the sustainable development discourse. Markunsen (2003) has put forth the following arguments to elucidate why it is imperative to consider religion/faith influences in the sustainable development discourse. First, mainstream environmental policies are proving insufficient to deal with the harrowing impacts of environmental degradation such as growing level of biodiversity losses, global warming, and increasing level of air/water pollution. This is causing increasing level of discontentment and disenchantment among individuals and social groups and highlighting the urgency of developing new paradigmatic approaches to resolve this crisis. Secondly, contemporary environmental challenges are multidimensional in nature. They not only comprise of an economic prerogative but are concerned with issues of ethics, morality, and social justice. Consequently, ecological challenges cannot be treated in a separate silo and distilled from larger socioeconomic challenges of economic inequality, social and environmental injustices, and global governance. Fourth, civil society actors comprising of nongovernment organizations, spiritual/religious leaders, grassroot organizations, and academics can play a leading role in "humanizing" existing ecological discourses and advancing collective solutions to sustainability challenges rooted in the notion of "interconnectedness of all forms of humanity" that remains common to most spiritual/faith traditions.

1.3 Role of Faith Traditions in the Sustainable Development Discourse

Based on our review of literature in the previous sections, we comprehend that faith traditions can contribute to existing discourses on sustainable development in a three-pronged way. First, most faith traditions incorporate a plethora of humanistic and ethical values such as kindness, courage, honesty, truthfulness, integrity, and community which can be differentially interpreted by individual believers of the faith as well as secular individuals and organizations to inform their sustainability practice (Narayanan, 2013). Second, the values embedded in various faith traditions can foster new forms of ground-level environmental activism in differing contexts. Third, faith traditions can foster pro-environmental attitudes and behaviors in individuals such as compassion for the natural environment, mindfulness, frugality, and sustainable consumption which can then translate into wider sustainability initiatives and also improve the proclivity of firms and organizations for pro-environmental actions. We elucidate upon each of these aspects in the following sub-sections.

1.3.1 Incorporation of Value Systems in the Sustainable Development Discourse

It is being increasingly recognized that creative solutions to the environmental crisis would require both individuals and organizations to take a "leap of faith" and commit to ethical/moral values and principles (Vu et al., 2022, p. 5). Religion/faith can emerge as an important source of such value systems. Historically, moral values and beliefs tend to be strongly embedded in faith traditions. This is particularly so in contexts where religion remains a strong macro-social force. In this context, Seiple (2016, p. 5) remarked: "If 84% of the world population believe in something greater than itself, it stands to reason that the factors will influence and impact global systemic challenges. Faith communities can act as a moral bulwark against the corruption that comes with a lack of legal framework. This capacity to act as a moral reference point is essential to the conversation about what a global society wants to become. Faith communities are also de facto trust networks. They can transcend divides, validating or invalidating different approaches to the common good. As such, they are also natural distribution networks."

There are several examples of how faith traditions and faith leaders are steering the sustainable development discourse by bringing in the influence of religious and moral values. In 2015, Pope Francis' papal encyclical revoked the moral imperative to provide stewardship to all forms of nature and gave a clarion call to Christian communities to work toward prevention of climate change, alleviation of poverty, and protection of vulnerable sections of the society (Francis & McDonagh, 2016). In Thailand, Buddhist traditions were drawn on to create a model of sufficiency economy, amid the Asian financial crisis of 1997/1998 (Song, 2020). Sufficiency economy ascribes to the Buddhist concept of "middle path" – avoidance of extremes of sensual pleasures on one hand and ascetic denial on the other hand (Vu & Burton, 2021). The idea of a sufficiency economy adopts a step-by-step approach to sustainable development centered on three key components – moderation, reasonableness, and prudence – and supported by values of knowledge and virtue (Mongsawad, 2010). In an empirical study of small and medium enterprises across seven countries, Abdelzaher and Abdelzaher (2017) illustrated how Muslim entrepreneurs leveraged Islamic values such as the notion of "humans as trustees of the natural environment," "protection of natural environment as a form of worship," and "environment as a source of blessings for humans" to create new forms of sustainable business practices in their respective organizational settings. The key findings of the study suggested how Islamic values inspired these entrepreneurs to adopt responsible/sustainable business practices as a moral and ethical imperative, even as they worked in institutional contexts characterized by weak environmental regulations. Singh (2022 b) illustrates how the Sikh concept of "seva," or voluntary service, is being employed by organizations and individuals in the Punjab state of India to create community-based models of renewable energy production, rooted in indigenous resources and technologies. Grzeda (2019) analyzed how the Jewish principle of *Tikkun Olam* (repairing and healing) can be employed to create new forms of conscious capitalism, embedded in spiritual values departing from conventional concepts of business philanthropy and corporate social responsibility.

These examples highlight how religious/faith traditions can provide a moral and ethical compass to individuals to work toward environmental protection. These can translate into local, grassroot level initiatives which represent a good fit with UN's sustainable development goals.

1.3.2 New Forms of Ecological Activism

Religious/spiritual can promote new forms of environmental and social activism through grass root level initiatives. The most famous of these constitutes the famous *Chipko* movement of the 1980s in the Himalayan belt of India. This movement was inspired by the Gandhian principle of *satyagraha* or nonviolent social action. *Chipko* means "cling to" or "hug" and involved community-based vigils to guard against the felling of trees. Participants in the Chipko movement advocated for forests as self-renewing systems, rather than economic resources for short-term economic gains. The Chipko movement was centered on three basic principles: (a) not to cut trees, (b) preservation of natural environment, and (c) being merciful to the natural environment (Shiva, 1988). Christian organizations have launched grassroot movements like Creation Care and Evangelical Environmental Network that inspire individuals to inculcate pro-environment attitudes and make lifestyle changes in areas of travel, consumption, food habits, and community engagement by employing the Christian concept of "providing stewardship to all of God's creation" (Clements et al., 2014, p. 86). Sikh organizations in India have launched a grassroot level "Green *Gurdwara* [temple]" campaign which involves tree plantation drives, encouragement to bioenergy projects, impetus to organic farming, and making holidays/processions more environmentally friendly (Singh, 2022b).

There have also been numerous inter-faith initiatives and dialogues to bring together faith leaders and practitioner communities to tackle sustainability challenges in a systematic manner. For instance, the Green Pilgrimage Network (GPN) launched in 2011 endeavored to make major pilgrimage cities across the world more sustainable (Prill, 2015). The GPN incorporates 20 major pilgrimage cities including Assisi, Jerusalem, Bethlehem, Varanasi, Nanded, and Amritsar (Prill, 2015). The GPN promotes sustainable tourism, tree plantation, greening of faith festivals, recycling waste, water purification, and protection of wildlife (GPN, n.d.). In 2007, the United Nations Development Programme partnered with the Alliance for Religion and Conservation[1] (ARC) to work with major faiths to address issues of global warming and climate change through the development of faith-based "long-term commitment of a living planet" (Alliance for Religions and Conservation, n.d.). This network brought together faith leaders from eight major world religions: Baha'ism, Christianity, Daoism, Hinduism, Islam, Judaism, Shintoism, and Sikhism and epitomizes the "largest civil society movement in climate change history" (Guardian, 2009). The commitments made in ARC led to a number of grassroot initiatives in different geographical contexts. For instance, a massive tree plantation drive in Tanzania by the Diocese of the Evangelical Lutheran Church of Tanzania. Sikh Gurdwaras in India started using sustainable energy sources for preparation of food *(langar)* in their community kitchens. These community kitchens feed approximately 30 million people every single day. Jewish participants advocated for converting *Shabbat* into an environmental celebration of avoiding consumption (Guardian, 2009).

[1] Alliance for Religion and Conservation (ARC) was a UK-based organization founded by Prince Philip in 1995. This organization aimed to help major religions of the world develop environmental programs based on their core values, teachings, and beliefs. Martin Palmer, the secretary general of the ARC, announced its closure in June 2019.

1.3.3 Potential for Self-Development and Inculcation of Pro-environmental Attitudes

Religion and faith-based approaches also offer opportunities for personal self-development and inculcation of pro-environmental attitudes and behaviors. As we fast approach a state of planetary emergency, there is heightened impetus on individual responsibility and cultivation of pro-environmental attitudes and behaviors (Bauer & Menrad, 2019). In this context, Hamilton (2008, p. xi–xii) writes:

> In an age of over-consumption, intemperance and moral confusion, the structures that prevent us from flourishing have lodged themselves in our psyche. The source for the kind of transformation that is now needed lies beyond the cultural, political and social philosophies that have formed the bedrock of progressive thought. We need to look to metaphysics – ideas about knowing and being that are beyond the psychological and social structures that condition everyday experience – to discover what unites us all in our humanity.

Existing studies associate religious individuals with values such as heighted self-awareness (McCullough & Willoughby, 2009) and self-effacement (Lefkowitz, 2017) which in turn can lead to cultivation of more ecologically friendly attitudes and lifestyles. In recent years, a number of studies have highlighted how individual environmental consciousnesses can provide intrinsic motivation for lifestyle changes and more sustainable patterns of consumption in society (Fehr & Schmidt, 2006; Brekke & Johansson-Stenman, 2008; Stern, 2008; Knetsch, 2010; Carlsson & Johansson-Stenman, 2012; Scales, 2014). Given the primacy of spiritual/religious influences among a large section of the global population, religious influences can have a strong influence of ethical values and judgments related to natural environment. This has been empirically investigated by some studies as well. For instance, Pace (2013) examined how Buddhist values such as compassion, kindness, and equanimity helped to foster anti-consumerist attitudes among the surveyed population. In another study, Felix and Braunsberger (2016) conducted a large market survey in North Mexico and employed structural equation modelling to assess people's proclivity to buy green products. The key results from this study demonstrated that participants with intrinsic religious orientations displayed a greater propensity to buy green products, compared to nonreligious participations.

In spite of the varied and multiple influences of faith-based approaches to contemporary sustainability problems, there has been limited scholarship in this area. In this context, Kirton and Greene (2021) describe how spiritual influences remain a mere "organizational catchphrase" that are rarely incorporated in ground-level organizational/management studies. This work endeavors to contribute to the emerging body of research in this critical area by presenting how the value systems and moral principles enshrined in faith traditions can help to develop new conceptual models of sustainable development, in the face of the devastating impacts of climate change crisis.

Following the introductory chapter, the volume brings together the ecological traditions of major world religions including Christianity, Buddhism, Confucianism, Judaism, and Sikhism, as well as underrepresented faith traditions such as the Order of Saint Benedict and Bahaism in ten subsequent chapters. The various contributory chapters in this volume use differing theoretical perspectives, interdisciplinary approaches, and methodological frameworks to advance existing scholarship on how the principles and values embedded in different faith traditions may provide the conceptual and practical basis for creation new sustainable development pathways. The contributions in this volume specially focus on the opportunities offered by faith-based approaches to sustainable development, how these may be replicated at the organizational level, as well as the contestations and tensions involved in their ground level implementation.

1.4 Summary of Chapters

This section presents a brief overview of the chapters included in the edited volume.

Chapter 2: "Guru Nanak's ecological legacy: New views and practices for sustainable development" by Nadia Singh explores the environmental, ethical, social, and cosmological aspects of Guru Nanak's writing (the first Sikh master) and how these provide the basis for creation of sustainable economies. The chapter identifies how Guru Nanak's writings provides insights into the environment-economy-society dynamic, the concept of material well-being, the drives of unsustainability in society, and their distinct sociocultural sub-texts. Nanak's worldview represents a model of a community driven needs-based society and illustrates how collective, social strategic action can modify the social fabric of human society and foster environmental and social sustainability in the long run. The final part of the chapter analyzes how superimposition of Guru Nanak's wisdom on comprehensive sustainability frameworks like the three pillars model can help us to formulate specific strategic actions and create a new blueprint for sustainable economic development in the long run.

Chapter 3: "Family firms: The impact of a Christian perspective on environmental protection and sustainability practices" by Allan Discua Cruz and M. Isabella Cavalcanti Junqueira employs a Christian perspective to analyze sustainability operations in family firms. The key findings of the study reveal how Christian family firms recognize the current environmental crisis as a moral, ethical, and faith-based issue and influences the decision-making principles and sustainability efforts of family firms. The analytical model developed in the chapter that a Christian-centered view of the environment results in six interrelated dimensions related to the sustainability of Christian family firms: humility, respect, selflessness, moderation, mindfulness, and responsibility.

Chapter 4: "Benedictine Perspective on Sustainability: Economic, Environmental and Social Inspiration for Organizations" by Sheila Hanson and Ksenia Keplinger presents a detailed treatise on the concepts of social, economic, and ecological sustainability enshrined in the Order of St Benedict. The chapter also underlines how Benedictine principles can give impetus to new forms of sustainability principles at the organizational level.

Chapter 5: "Confucianism and Sustainability" by Irene Chu uses the Confucian concept of harmony (*he*), correct action (*li*), and excellent person (*junzi*) to demonstrate how these three principles can be used to create "grand harmony" between environment, society, and individuals. The chapter goes on to illustrate how businesses can apply these principles not only to create and maintain a harmonious working environment internally but also with wider stakeholders and the natural environment.

Chapter 6: "Buddhist economics: Philosophical premises and environmental policy implications" by George Gotsis expands on the perspective inaugurated by scholars such as E.F. Schumacher, P.A. Payutto, and Shérab Tendar and analyzes how Buddhist economics can provide a humane and compassionate responses to persistent sustainability and environmental responsibility concerns facing humanity.

Chapter 7: "Cultivation of Loving-Kindness and Compassion: A Societal Solution to Uphold Strong Sustainability Principles in Ecological Policies" by Viet-Ngu Hoang, Tsewang Topden, and Florence Ingram presents an analysis based on the classical cognitive framework of Buddhism through the lenses of Middle Way Consequence philosophy. The authors present a narrative arguing that cultivating loving-kindness and compassion (LK&C) in individuals is key to the accumulation of altruistic capital, which can be a mental motivator for pro-environmental behavior and attitudes.

Chapter 8: "Calling to the Mind: Exploring the Interlinkages of Mindfulness, Morality and Ecological Sustainability" **by** Mai Chi Vu and Nadia Singh uses field-based research among Buddhist and Sikh mindfulness practitioners in India and Vietnam to demonstrate how ethical dimensions of mindfulness can help to motivate an orientation toward sustainability and other-centeredness. These in turn pay a "double dividend" in terms of contribution to a sustainable way of life as well as a greater sense of well-being.

Chapter 9: "Judaism and Ecological Discourse: What a Jewish Religious Perspective Offers to Contemporary Sustainability Dialogue" by Susan Case explores ways classical Jewish sources, Jewish law (*Halakha*), and the Hebrew Bible (*Tanakh*) can be analyzed around ecological themes contributing to global sustainability practices to protect our environment. Perspectives used focus on interconnecting humanity and nature linking environmental stewardship, protection, and the idea of repairing the world (*tikkun olam*). The chapter analyzes how Jewish perspective embedded in the principles of respect for human and nonhuman life and social justice offers insight into a new framework of "just sustainability," wherein environmental protection is combined with ecological justice. This framework can offer a new vision to individuals and organizations to tackle environmental issues.

Chapter 10: "A collective sustainability approach based on the Bahá'í principles" by Legha Momtazian proposes a "collective sustainability framework," drawn from the Bahá'í Faith that enables an effective integration of various stakeholders and proposes benefits for the environment and the society irrespective of the perceived differences. This sustainability framework encompasses the principles of justice for all, unity in diversity, unity between science and religion, high station of work, and cessation of war expenditure as its building blocks. It concludes that such a collective sustainability framework can develop through a value intersubjectivity approach, to reflect the context and environment it applies to and can be employed in sustainability agendas as well as in organizational policy decisions and sustainability initiatives.

References

Abdelzaher, D. M., & Abdelzaher, A. (2017). Beyond environmental regulations: Exploring the potential of "eco-Islam" in boosting environmental ethics within SMEs in Arab markets. *Journal of Business Ethics, 145*(2), 357–371.

Abeydeera, S., Kearins, K., & Tregidga, H. (2016). Buddhism, sustainability and organizational practices fertile ground? *The Journal of Corporate Citizenship, 61*, 44–70.

Alliance for Religion and Conservation. (n.d.). *ARC and the faiths*. http://www.arcworld.org/arc_and_the_faiths.html

Banerjee, S. B., & Bonnefous, A. M. (2011). Stakeholder management and sustainability strategies in the French nuclear industry. *Business Strategy and the Environment, 20*(2), 124–140.

Bauer, A., & Menrad, K. (2019). Standing up for the Paris agreement: Do global climate targets influence individuals' greenhouse gas emissions? *Environmental Science & Policy, 99*, 72–79.

Baumol, W. J., & Oates, W. E. (1988). The theory of environmental policy. Cambridge university press.

Bouckaert, L., & Zsolnai, L. (2012). Spirituality and business: An interdisciplinary overview. *Society and Economy, 34*(3), 489–514.

Brekke, K. A., & Johansson-Stenman, O. (2008). The behavioural economics of climate change. *Oxford Review of Economic Policy, 24*(2), 280–297.

Briassoulis, H. (1999). Who plans whose sustainability? Alternative roles for planners. *Journal of Environmental Planning and Management, 42*(6), 889–902.

Carlsson, F., & Johansson-Stenman, O. (2012). Behavioral economics and environmental policy. *Annual Review of Resource Economics, 4*(1), 75–99.

Clements, J. M., McCright, A. M., & Xiao, C. (2014). Green Christians? An empirical examination of environmental concern within the US general public. *Organization & Environment, 27*(1), 85–102.

Daniels, P. (2007). Buddhism and the transformation to sustainable economies. *Society and Economy, 29*(2), 155–180.

Daniels, P. L. (2010). Climate change, economics, and Buddhism—Part 2: New views and practices for sustainable world economies. *Ecological Economics, 69*(5), 962–972.

Ehrenfeld, J. R., & Hoffman, A. J. (2013). *Flourishing: A frank conversation about sustainability*. Stanford University Press.

Fehr, E., & Schmidt, K. M. (2006). The economics of fairness, reciprocity and altruism–experimental evidence and new theories. In *Handbook of the economics of giving, altruism and reciprocity* (Vol. 1, pp. 615–691).

Felix, R., & Braunsberger, K. (2016). I believe therefore I care: The relationship between religiosity, environmental attitudes, and green product purchase in Mexico. *International Marketing Review, 33*(1), 137–155.

Francis, P., & McDonagh, S. (2016). *On care for our common home, Laudato Si': The encyclical of Pope Francis on the environment*. Orbis Books.

Gottlieb, R. S. (Ed.). (2010). *Religion and the environment* (Vol. 1). Routledge.

GPN. (n.d.). *Green Pilgrimage network*. http://www.arcworld.org/projectsa1c5.html?projectID=629

Grzeda, M. (2019). Tikkun Olam: Exploring a spiritual path to sustainability. *Journal of Management, Spirituality & Religion, 16*(5), 413–427.

Guardian. (2009). *Impact of religion will have "deeper roots" than Copenhagen*. https://www.theguardian.com/environment/2009/nov/02/impact-religion-copenhagen

Hallin, A., Karrbom-Gustavsson, T., & Dobers, P. (2021). Transition towards and of sustainability—Understanding sustainability as performative. *Business Strategy and the Environment, 30*(4), 1948–1957.

Hamilton, C. (2008). *The freedom paradox: Towards a post-secular ethics*. Allen & Unwin.

Intezari, A. (2015). Integrating wisdom and sustainability: Dealing with instability. *Business Strategy and the Environment, 24*(7), 617–627.

Jenkins, W., & Chapple, C. K. (2011). Religion and environment. *Annual Review of Environment and Resources, 36*(1), 441–463.

Johnston, L. F. (2014a). Sustainability as a global faith? The religious dimensions of sustainability and personal risk. *Journal of the American Academy of Religion, 82*(1), 47–69.

Johnston, L. F. (2014b). *Religion and sustainability: Social movements and the politics of the environment*. Routledge.

Johnston, P., Everard, M., Santillo, D., & Robèrt, K. H. (2007). Reclaiming the definition of sustainability. *Environmental Science and Pollution Research International, 14*(1), 60–66.

Kirton, G., & Greene, A. M. (2021). *The dynamics of managing diversity and inclusion: A critical approach*. Routledge.

Knetsch, J. L. (2010). Values of gains and losses: reference states and choice of measure. *Environmental and Resource Economics, 46*, 179–188.

Landrum, N. E., & Ohsowski, B. (2018). Identifying worldviews on corporate sustainability: A content analysis of corporate sustainability reports. *Business Strategy and the Environment, 27*(1), 128–151.

Lefkowitz, J. (2017). *Ethics and values in industrial-organizational psychology*. Routledge.

Markusen, A. (2003). Fuzzy concepts, scanty evidence, policy distance: The case for rigour and policy relevance in critical regional studies. *Regional Studies, 37*(6–7), 701–717.

Marques, J. (2010). Toward greater consciousness in the 21st century workplace: How Buddhist practices fit in. *Journal of Business Ethics, 92*(2), 211–225.

McCullough, M. E., & Willoughby, B. L. (2009). Religion, self-regulation, and self-control: Associations, explanations, and implications. *Psychological Bulletin, 135*(1), 69.

Mongsawad, P. (2010). The philosophy of the sufficiency economy: A contribution to the theory of development. *Asia Pacific Development Journal, 17*(1), 123.

Munn, R. E. (Ed.). (2002). *Encyclopedia of global environmental change*. Wiley.

Narayanan, Y. (2013). Religion and sustainable development: Analysing the connections. *Sustainable Development, 21*(2), 131–139.

Pace, S. (2013). Does religion affect the materialism of consumers? An empirical investigation of Buddhist ethics and the resistance of the self. *Journal of Business Ethics, 112*(1), 25–46.

Pew Research Centre. (2017). *Global religious landscape*. https://www.pewresearch.org/religion/2012/12/18/global-religious-landscape-exec/

Prill, S. E. (2015). Sikhi and sustainability: Sikh approaches to environmental advocacy. *Sikh Formations, 11*(1–2), 223–242.

Rakodi, C. (2012). Religion and development: Subjecting religious perceptions and organisations to scrutiny. *Development in Practice, 22*(5–6), 621–633.

Sandefur, R. L., & Laumann, E. O. (2009). A paradigm for social capital. In *Knowledge and social capital: Foundations and applications* (pp. 69–87). Routledge.

Scales, I. R. (2014). Green consumption, ecolabelling and capitalism's environmental limits. *Geography Compass, 8*(7), 477–489.

Schwab, K., & Zahidi, S. (2020). *Global competitiveness report: Special edition 2020*. World Economic Forum.

Seiple, C. (2016). What faith can do for global systemic challenges. In *World economic forum, ed., the role of faith in systemic global challenges. Geneva: World economic forum* (pp. 5–6).

Shiva, V. (1988). *Staying alive: Women, ecology, and survival in India* (Vol. 84). Kali for Women.

Singh, N. (2022a). Bioenergy and Ecosocialism. In *The Routledge handbook on ecosocialism* (pp. 320–327). Routledge.
Singh, N. (2022b). Sikhism and sustainability: New approaches to environmental ethics. In *Faith traditions and practices in the workplace volume I* (pp. 37–61). Palgrave Macmillan.
Singh, N. (2022c). Sikhism and Covid-19: Ethics of community service and activism. *Sikh Formations*, 1–13.
Singh, K., & Clark, J. S. (Eds.). (2016). *Voices from religions on sustainable development*. BMZ.
Singh, P., & Singh, N. (2019). Political economy of bioenergy transitions in developing countries: A case study of Punjab, India. *World Development, 124*, 104630.
Song, H. C. (2020). Sufficiency economy philosophy: Buddhism-based sustainability framework in Thailand. *Business Strategy and the Environment, 29*(8), 2995–3005.
Starik, M., & Rands, G. P. (1995). Weaving an integrated web: Multilevel and multisystem perspectives of ecologically sustainable organizations. *Academy of Management Review, 20*(4), 908–935.
Stern, N. (2008). The economics of climate change. *American Economic Review, 98*(2), 1–37.
Summer, J. (2005). *Sustainability and the civil commons: Rural communities in the age of globalization*. University of Toronto Press.
Vu, M. C., & Burton, N. (2021). Bring your non-self to work? The interaction between self-decentralization and moral reasoning. *Journal of Business Ethics, 181*(2), 427–449.
Vu, M. C., Singh, N., Burton, N., & Chu, I. (2022). Introduction: Faith traditions in unprecedented times. In *Faith traditions and practices in the workplace volume I* (pp. 3–14). Palgrave Macmillan.
WCED, S. W. S. (1987). World commission on environment and development. *Our Common Future, 17*(1), 1–91.
Welford, R. J. (1998). Corporate environmental management, technology, and sustainable development: Postmodern perspectives and the need for a critical research agenda. *Business Strategy and the Environment, 7*(1), 1–12.
Will, J. A., & Cochran, J. K. (1995). God helps those who help themselves. The effects of religious affiliation, religiosity, and deservedness on generosity toward the poor. *Sociology of Religion, 56*(3), 327–338.
WWF. (2022). *How can you help tackle climate change.* https://www.wwf.org.uk/climate-change-and-global-warming

Part II
Sikhism

Chapter 2
Guru Nanak's Ecological Legacy: New Views and Practices for Sustainable Development

Nadia Singh

Abstract In recent years, the global economy has been suffering from multi-pronged disturbances characterized by increasing economic turbulence, high levels of environmental degradation and sociocultural upheavals, rise of hyper-nationalism, misinformation, and extremism. The ecological crisis confronting humanity is especially grave, as we have reached a "tipping point" in case of many key natural resources. It is being increasingly recognized that the existing approaches to solving the environmental crisis are piecemeal and incremental. These approaches have largely failed to have the desired impact. Therefore, it is crucial to explore new worldviews and philosophical paradigms to charter a sustainable development pathway. In this context some scholars are trying to reinterpret religious texts and incorporate insights from major world religions in mainstream sustainability frameworks. This chapter endeavors to contribute to this critical of research by exploring the ecological, ethical, social, and cosmological aspects of Guru Nanak's writings and analyzing how these can help to create a new vision for sustainable development.

Keywords Guru Nanak · Ecology · Sustainable development · Cosmology · Religion

2.1 Introduction

Rapid environmental degradation constitutes the gravest crisis facing humanity today. Many scholars have reflected on the varied ways in which the ecological crisis threatens the very survival of our planet (Barnosky & Hadly, 2016; Lamentowicz et al., 2019; Singh, 2022a). Lenton et al. (2008) gave the concept of a

N. Singh (✉)
Newcastle Business School, Northumbria University, Newcastle upon Tyne, UK
e-mail: nadia.singh@northumbria.ac.uk

N. Singh et al. (eds.), *Faith Traditions and Sustainability*, Management, Change, Strategy and Positive Leadership, https://doi.org/10.1007/978-3-031-41245-5_2

"tipping point," as a threshold level at which the most miniscule disturbances could lead to critical changes in the level of our natural resource base. In a subsequent study, Lenton et al. (2019) analyzed how some key natural resources such as Amazon rainforests and artic ice sheets have already reached a "tipping point" and remain close to extinction. The authors hypothesized that given the close integration of all forms of biosphere, these changes may lead to irreversible damages in future years. The Living Planet Report (2022) also points out that there has been a 69 percent decline in monitored wildlife populations and an 83 percent decline in freshwater populations between the years 1970 and 2018. In the year 2020, the Alliance of World Scientists comprising a group of 11,000 scientists from across 153 countries of the world presented data on key ecological parameters and warned how human activities are creating a state of "climate change emergency" with excessive emission of greenhouse gas emissions continued use of fossil fuel and unsustainable patterns of production and consumption in modern societies. The *Provisional State of Global Climate in 2022* published by the World Meteorological Association reveals that the average global temperatures during the period 2013–2022 are 1.05–1.27 °C higher than the baseline levels, indicative of severe long-term trends in global warming (World Meteorological Association, 2022).

The catastrophic impacts of anthropogenic climate change reveals that we cannot continue with the "business as usual" approach and must make drastic changes in the nature and direction of environmental policies. In recent years, a number of scholars have questioned the efficacy of neoclassical economic policies in dealing with the gravity of the ecological crisis (Jacobs, 2012; Janicke, 2012; Dawson, 2013; Singh & Singh, 2019; Singh, 2022a). At the outset, mainstream economists denied the gravity of ecological problems. For instance, Milton Friedman ridiculed the idea of climate change and remarked that horses produce more pollution than humans (cited in Castro, 2004, p. 203). Robert Solow described the sustainable development as a "glib idea" which is "undesirable" and "unfeasible" and akin to being "asked to play Peter Pan and flapping one's wings" (Solow, 1993, p. 180). However, over the years as environmental degradation continued to ravage the global economy, mainstream neoclassical economists begin to consider nature as "form of capital" or "input in the process of production" (Singh, 2022b), which is readily substituted with other forms of human-made and technological capital to continue the process of production, material accumulation, and subsequent economic growth (Singh & Singh, 2019; Singh, 2022b). They recognize that there is a stock of "critical" natural capital, which must be maintained at fixed levels in order to continue the process of economic growth unabated (Castro, 2004). These economists contend that this stock of critical natural capital may be maintained through market-based mechanisms (Baumol & Oates, 1988) such as carbon taxes, corporate subsidies, tradeable pollution permits, and assigning property rights to environmental goods and services.

These approaches are being critiqued as "too limited, disjointed and weak, and incapable of inducing the fundamental changes required to arrest anthropocentric climate change" (Daniels, 2010, p. 359). Other scholars have described mainstream environmental policies as "reductionist" in nature (Singh & Singh, 2019). These approaches tend to consider humans as merely rational individuals (Castro, 2004; Adman & Ozkaynak, 2002) who are driven exclusively by an economic prerogative. These approaches do not consider the sociological, cultural, political, and psychological factors which influence an individual's relationship with society and the natural environment. Given the glaring limitations of mainstream environmental approaches, new philosophical paradigms and theoretical approaches are being added to the tapestry of existing sustainability frameworks to distill the influence of factors such as gender, socioeconomic inequities, poverty, and behavioral factors on the drivers of environmental degradation and effective adaptation strategies.

An important theoretical development in this regard has been with respect to incorporation of the ethical principles and value systems embedded in religious/faith traditions into existing sustainability frameworks (Singh, 2022a). It is being proposed that faith-based frameworks can help to create new pathways for tackling ecological challenges, especially in the contexts where religion is a dominant sociocultural influence in society (Johnston, 2014; Singh, 2022a). These faith-based approaches to tackle issues of environmental degradation depart from mainstream economics frameworks that have dominated environmental policymaking so far by bringing in ethical principles and moral nuances such as "honesty," "responsibility," "social justice," and "integrity" into the discussion of sustainable development paradigms.

Some scholars are attempting to create environmental sensitivity among people through novel interpretation of principles and values embedded in various religious traditions (Prill, 2015; Daniels, 2007, 2010). For instance, Helfaya et al. (2018) analyzed how the principles enshrined in the Qur'an such as "humans as gatekeepers of the natural environment" and "nature as a holy concept" can be employed to educate followers of Islam on the dynamic interlinkages between the human world and the natural world and promote a heightened level of ecological consciousness. In another study Daniels (2011) argues how Buddhist doctrine of "karmic" merit can enable effective evaluation of the environmental footprint of production and consumption activities in modern economies and help in effective transition to sustainable production/consumption. Jenkins (2013) analyzes how the Christian traditions of grace and salvation can help to create new forms of environmental activism, rooted in stewardship of nature and ecological justice. In other studies Ashton (2014) and Khaleel and Shobha (2021) analyze how the Hindu ethic of *ahimsa* (nonviolence) and *dharma* (righteous action) can be extended to the domain of the natural environment to cultivate an ethics of protection of the natural environment. In recent years, there have also been formation of inter-faith partnerships and networks in an attempt to bring together faith leaders and religious communities across the world. The Green Pilgrimage Network (GPN) was formed in 2011 to promote "a positive environmental footprint of pilgrim activities" (EPGN, n.d.). The members of GPN comprised of 28 holy cities from different parts of the world including Amritsar (Sikhs), Assisi (Roman Catholics), Haifa (Bahai), and Louguan (Daoists)

(EPGN, n.d.). In 1986, the Alliance for Religion and Conservation ARC, n.d.[1] was created in Assisi, Italy, to promote an understanding of how partnerships between religious leaders and secular environmental movements could gave a fillip to the environmental/green movement (ARC, n.d.).

In consonance with other religious communities, Sikhism is increasingly adopting an ecological focus and devising new strategies and pathways to encounter contemporary ecological challenges facing humanity. This process has accelerated noticeably since Sikh temples (*Gurdwaras*) started participating in a *Green Gurdwara* movement through measures like preparation of organic food in the community kitchen, adoption of renewable energy, tree plantation drives, and creating community awareness on ecological issues (Singh, 2022a). Sikhism was also a part of ARC among nine other major world religions. The Sikh Faith Statement said:

> Sikhism opposes the idea that the struggle of the human race is against nature and that human supremacy lies in the notion of "harnessing" nature. The objective is harmony with the eternal—God—which implies a life of harmony with all existence. Striving for a life of harmony, therefore, also implies a life of supporting individual rights and environmentalism—a life that works against injustice toward anybody and anything. (ARC, Sikh Faith Statement, n.d.-b)

Sikh environmentalism has however received little scholarly attention so far, even though Sikhism as a lived tradition can help in creating a new paradigm for tackling contemporary environmental challenges. In this paper I attempt to explore this under-researched area, by examining the socioecological dimensions of Guru Nanak's writings, the founder of the Sikh faith. I analyze how his life, work, and teachings can enhance our understanding of the current drivers and factors influencing environmental degradation in society. The chapter departs from analyzing Guru Nanak's philosophy as merely a religious exposition and instead presents the discussion as a source of new insights for bringing about a socioeconomic transformation to sustainability.

2.2 Guru Nanak: Life and Philosophy

Guru Nanak (1469–1539) was the founder of the Sikh faith. He was born in Talwandi Rai Bhoe (present-day Nankana Sahib, Pakistan). His father, Mehta Kalian Das Bedi, was an accountant in the village. There are various tales and folklores associated with the life of the young Nanak. Singh (2004, p. 28) describes how, by the age of 5, Nanak began to question the very purpose of life. He took little interest in formal studies and often sought the company of mystics and holy men, conversing

---

[1] The ARC initially brought together leaders of five main religious groups: Buddhism, Christianity, Hinduism, Islam, and Judaism at Assisi, Italy. This place was chosen because it was the birthplace of St Francis, the Catholic saint of ecology. In a subsequent conference in 1995 at the Windsor Castle, England, representatives of nine major religious came together and the ARC was formerly launched (ARC, n.d.).

on the philosophies of life. From an early age, Nanak rejected ritualism and dogmatic practices. At the age of 11, he refused to participate in a ceremony where a pandit would tie a *janeu* (sacred thread) around him, sanctifying his superior status in India's rigid caste hierarchy. Even at that age, Nanak had come to the conviction that a person's merit should not be dictated by a person's caste identities, which merely reflect an accident of birth. Although there are numerous anecdotes of the divine potentialities of the young Nanak, he is said to have attained spiritual enlightenment at the age of 30, around 1499 (Singh, 2004, p. 29). For 3 days, Nanak disappeared, leaving his clothes by the bank of a stream called Kali Bein in Sultanpur Lodhi. He reappeared on the fourth day and proclaimed, "There is no Hindu, there is no *Mussalman*" (Singh, 2004, p. 31). This period marks the end of the first phase of his life and his quest for eternal truth (Singh, 2004, p. 31).

In the second phase of his life, Nanak started traveling extensively to different parts of India and abroad to spread his divine message as far as possible. His travels extended from Ladakh in the north to Ceylon (present-day Sri Lanka) in the south. He toured from Baghdad in the west to Jagannath puri along the eastern course of India (Singh, 2004, p. 31). The *janamsakhis* contain numerous accounts of Nanak's long travels or *uddasis* and how he challenged superstitions and orthodox religious practices through practical examples. During these travels he adopted an "outlandish garb" that "combined the styles of Muslim fakirs and Hindu sadhus" (Singh, 2004, p. 32). This often-confused people and caused them to question, "Art thou a Hindu or a *Mussalman*?"

Guru Nanak's philosophy was rooted in the idea of monotheism. The central praxis of his philosophy was the concept of oneness of humanity (*ik onkar).* In his teachings, he not only reflected on spiritual concepts but also gave a blueprint of the principles of rightful conduct and accepted social behaviors which Sikh followers should live by. He developed a novel social and political platform rooted in the principles of equality, social justice, rightful conduct, and truthfulness. He summarized his religious values in terms of three everyday practices – *Wand Chakna* (sharing with others the fruits of one's labor and giving to those who are less fortunate), *kirat karo* (earning an honest livelihood through hard work and toil), and *naam japna* (meditating). His teachings had deep sociocultural implications. He challenged the caste system and gender discrimination inherent in Hinduism and emphasized that external forms of worship such as ritualism, fasting, and idols are of little importance. His philosophy gave paramount importance to one's inner spiritual awakening.

2.3 Guru Nanak and Three Pillars of Sustainability

Guru Nanak's philosophy does not merely represent an article of faith. Rather, his writings reflect on human life and the environment based on rationality and scientific logic. Guru Nanak's worldview has been described as a "way of life" (Tatla, 2008), rooted in ethical values and morality, which would bring about both individual and societal well-being. Nature occupies an important place in Guru Nanak's philosophical paradigm. According to one scholar, Guru Nanak's writings represent an "eco-sophical" tradition, which "locates the socio-economic, and environmental problems of his time within the social relations of domination in society" (Lourdunathan, 2012, p. 12).

Eco-sophia represents a philosophical paradigm which accords primacy to achieving harmony between the human and the nonhuman world, through rejection of anthropocentrism[2] (Lourdunathan, 2012). Eco-sophism remains embedded in several Asian traditions such as Buddhism, Taoism, and Confucianism. In contrast to European individualism, these traditions espouse anti-individualism and provide inspiration to individuals to live in harmony with all forms of nature (Tinnell, 2012). However, Guru Nanak's philosophy differs from these traditions in two fundamental ways. First, in contrast to other philosophies that deny worldliness, Sikh affirms that the human world is itself representative of divinity. Guru Nanak's teaching emphasizes that we should strive to transform the world (earth) into a place of righteousness and purity, which should be conserved through honest labor, sharing of resources within the community, and rejection of greed/materialism. Second, Sikhism historically represents a tradition of class struggles against all forms of societal inequities and injustices. Within Sikh traditions, ecological pollution is conceptualized to include all forms of societal discrimination including those of gender, caste, and class. *Sangat*, the basic Sikh community, represents a society of equals and may be viewed as an eco-sophical community, reflective of harmony with nature and all its creations. These two traditions of sangat and class struggle within Sikhism can provide motivation for community-based environmental activism to solve contemporary ecological challenges. For instance, Singh (2022a) presents case studies of how Sikh individuals and organizations are creating inclusive models of solar and biomass energy in Punjab, India, on the basis of indigenous technologies, readily available feedstock, and participation of local rural communities. These models are distinct from the mainstream technocratic and market-oriented models of renewable energy development being promoted by large corporates, as they involve a "bottoms up approach" with the involvement of ground level stakeholders to build a common consensus and acceptability of these projects at the ground level.

In this context, Guru Nanak's socioeconomic and ecological teachings offer a good fit with the concept of the "three pillars of sustainability" (Purvis et al., 2019) approach – social well-being, economic upliftment, and environmental protection – that is being used in the contemporary global ecological discourse. Sustainability remains a multifaceted concept, conceptualized differently by different schools of thought. However, in Guru Nanak's philosophy, there remains a strong focus on its community-driven aspects, as well its integration with economic well-being and society welfare. Therefore, we conceptualize sustainability as a set of practices to

[2] Anthropocentrism is an idea which accords to primacy to human well-being over the well-being of other forms of nature.

encourage a community to make judicious use of available natural resources in order to strike equilibrium between the conflicting goals of economic upliftment, environmental protection, and social well-being.

The three-pillar framework adopted here deviates from standard economic theory that regards sustainability as continued economic growth (Daniels, 2007) while relegating social and ecological sustainability to merely precondition for economic growth (Daniels, 2007). In contrast, this framework advocates a holistic approach in where in societal, economic, and environmental aspects are all equally crucial for the achievement of a sustainable development trajectory and must work in tandem with each other for the survival and well-being of humanity as a whole.

The United Nations Framework Convention on Climate Change (UNFCCC) has recognized that the three pillars of social, economic, and environmental sustainability are crucial ingredients to the creation of environmentally friendly societies (The Tribune, 2019). Within this framework, sustainable development is defined as "a community's prudent use of natural resources to minimize the ecological footprint, while taking into account the equitable distribution of these resources, so that the present and future generations can attain a high degree of economic security, while maintaining the integrity of the ecosystem, upon which all life and production depends" (Neven et al., 2015) (Fig. 2.1).

Although the three-pillar framework has helped to develop a comprehensive framework to understand the drivers of environmental destruction, it suffers from two fundamental weaknesses (Daniels, 2007). *First*, the framework does not recognize that the various parameters of sustainable development remain "contextual" in nature. Hence, it does not elaborate upon the sociocultural dimensions of sustainability. These include the role of values, beliefs, as well as ethical, cultural

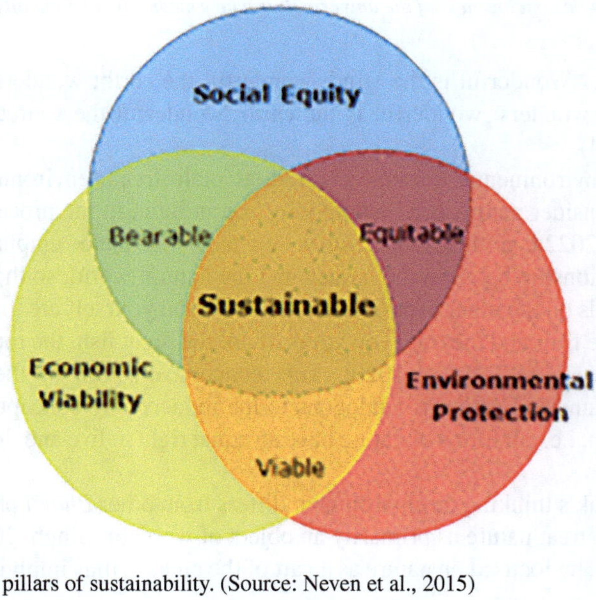

Fig. 2.1 Three pillars of sustainability. (Source: Neven et al., 2015)

dimensions, which can have an important mediating role in encouraging pro-environmental attitudes and behaviors. *Second*, the concept of well-being has not been clearly articulated within the three-pillar framework. Mainstream economists view well-being through a narrow prism of wealth accumulation and high levels of economic growth (Cropper & Oates, 1992; Hanley et al., 1997; Henisz, 2000; Tiba & Omri, 2017). However, the well-being of an individual is influenced by a range of factors including an individuals' psychology, mental and physical health, and personal and community relations. These factors have a profound influence on an individual's relationship with society and the natural environment (Daniels, 2010). These factors have not populated the three-pillar framework so far.

In this context, the philosophical paradigm espoused by Guru Nanak can provide a holistic framework to integrate the social, economic, and ecological aspects of sustainability, bringing in the role of sociocultural factors, as well as a deeper knowledge of the influences of well-being of the population. The next sections of the paper systematically analyze Guru Nanak's worldview on the environment, pollution, material well-being, and sustainable societies.

2.4 Guru Nanak's Understanding of the Natural Environment

Guru Nanak's writings elucidate the fundamental importance of nature on human and societal well-being. According to Lourdunathan (2012):

> *Nanak's philosophy represents the kind of wisdom that is suggestive of norms, ways of perceiving Nature, as against dominant world-views. It goes beyond anthropocentrism; towards holistic conciousness of the universe. It is a new mediation of humanity and society with Nature.*

Nanak wrote, "Wonderful is the wind; wonderful the earth; wonderful is the fire, which works wonders; wonderful is the earth, wonderful the source of creation" (SGGS, p.464).

Nanak's environmental doctrine contradicts mainstream environmental economists who consider nature as a "commodity" or an input in the process of production (Singh, 2022a, p. 40). Guru Nanak, on the other hand, emphasized on the intrinsic relationship between the human and the natural world, so that the survival of one depends on the other (The Tribune, 2019). "You yourself are the bumble bee, the flower, the fruit and the tree; you yourself are the great fish, the tortoise, and the cause of causes" (SGGS, p. 1020). This conceptualization of the relationship between humans and nature is analogous to the modern-day concept of biological egalitarianism, i.e., all forms of nature have an equal right to live and flourish (Singh, 2022a, p. 40).

Guru Nanak's thinking on environment differs from other *Bhakti* philosophers of the time, who treat nature as primarily an object of worship (Singh, 2022a). In contrast with this, he focused on nature as a part of the creator, thus implying that caring for nature and its preservation is essentially a form of worship (Singh, 2022a). He considered the world as a dynamic, interconnected web of relations, with no dividing line between human and nonhuman environment. He wrote, "Every creature in this world, every plant, the air, the water and fire, the sun, the moon and the earth are a manifestation of the creator" (SGGS, p. 1020). Several other passages in the Guru Granth Sahib elucidate on this relationship between nature and the divine. The most famous of these is:

Air, water and earth,
Of these are we made.
Air like the Guru's words gives the breath of life
To the babe born to the great mother earth
Sired by the waters
The day and night our nurses be
That watch us in our infancy
In their lap we play
The world is our playground (SGGS, p.8; translated by Singh, 2004, p. 328)

Guru Nanak's writings reveal that not only was he a nature lover; he also had a scientific conception of the natural environment. In the *Raag Maru Solhe*, Nanak talks about cosmology and how the world was essentially created out of nothingness. "For billions of years there was only utter darkness. There was no day and night, no moon and sun. The Lord alone sat in a profound trance. There was no air or water. There was no creation or destruction. There were no continents; neither regions, seven seas, river nor flowing water. There was neither birth nor death (SGGS: 1035). There was no Brahma nor Vishnu and Shiva." His writing challenged the commonly held belief at the time that a bull was holding the earth. Nanak went on to describe this bull as a "mythical creature" in the *Japji Sahib*. Prof. Hardev Singh Virk has proposed that Guru Nanak's ideas about the creation of the universe out of nothing are essentially akin to the Big Bang theory, proposed eons later by Stephen Hawking (Virk, 2018). He draws a parallel between Guru Nanak's vision of the emergence of the universe from a void, and Hawkin's scientific model, which propounds that the universe emerged from a singularity, preceding this event space and time did not exist. Moreover, according to both modern science and Guru Nanak's cosmological vision, there are innumerable galaxies in the universe, and in each of these, there are infinite number of solar systems containing stars and planets, all of which are in constant motion. This is exemplified in the following verses from the Sikh holy book, Sri Guru Granth Sahib:

There are planets, solar systems and galaxies. If one speaks of them, there is no limit, no end. There are worlds upon worlds of His Creation. As He commands, so they exist." (SGGS, p. 8)

I have searched and searched, across so many worlds, but without the Name, there is no peace. (SGGS, p. 255)

Many millions are the fields of creation and the galaxies (SGGS, p. 276)

Guru Nanak's life and teachings thus reflect the fact that environmental sustainability remains a necessary prerequisite for societal well-being and harmony. His writings reveal that we should go beyond the current form of "ecological fixing" and development of technocratic solutions to environmental problems. Instead, environmental protection should become a central tenet of social activity, as human and the nonhuman world cannot be compartmentalized. They are deeply intertwined through the process of social metabolism itself.

2.5 Guru Nanak's Conception of Pollution

Guru Nanak wrote in detail about pollution in society. However, his understanding of pollution was not limited to environmental pollution alone. He regarded all kinds of discrimination in society including untouchability and gender bias against women as forms of pollution. "He was a rational thinker who emphasized that all human beings are essentially composed of matter" (The Tribune, 2019). This is exemplified in the passage "Born out of flesh, in flesh does man live" (SGGS, pp. 1289–90, *Raga Malar*). He strongly condemned purification rituals espoused in the *Manusmriti* and said that the caste systems represent a systemic mechanism to perpetuate inequities, injustices, and exploitation on certain sections of the Indian society (The Tribune, 2019). In his teachings he focused on equality of all human beings and attempted to break the artificial barriers of caste, class, and gender divisions in Indian society. His egalitarian message emerges most powerfully in the manner in the phrase:

> *Among the low, let my caste be the meanest.*
> *Of the lowly, let me the lowliest be.*
> *O' Nanak, let such be the men I know.*
> *With such me let me keep company.*
> *Why should I try to emulate the great?* (SGGS, p.15, translated by Singh, 2004, p. 328).

Unlike the Bhakti scholars, whose anti-caste attitude was confined to the spiritual dimension alone, Guru Nanak challenged the caste system at the practical level as well (Takhar, 2011). Nanak's life is replete with numerous examples and stories of his commitment to uplift the so-called low-caste communities in India's rigid caste hierarchy. One such folklore concerns how Guru Nanak choose to stay in the house of a poor carpenter Bhai Lalo and rejected the hospitality of a wealthy landlord, Malik Bhago, whose riches had been won through deceit and exploitation of low-caste, working class people. Guru Nanak also preached against ritualism and superstitions, which were devoid of rationality and logical reasoning. He was especially critical of societal practices, which propagated discrimination against women. For instance, he condemned the notion of mensuration as a form of "pollution" and reproached those who regarded a garment stained with menstrual blood as polluted. "If clothes be stained with blood, the garment gets polluted. Who suck the blood of human beings; how can their mind be pure? Nanak, utter the Name of, from your mouth with sincere heart. Else are worldly ostentations, Man practices false deeds" (SGGS, p.140). On the other hand, he emphasized on the total equality of men and women.

From a woman, a man is born.
Without a woman, there can be none.
Nanak, only the one True Lord, is without a woman.
(SGGS, p. 473*).*

He scoffed at the hypocrisy of priests who emphasized on vegetarianism as a form of purity but did not hesitate to slaughter animals for sacrifice, "The fools argue about flesh and meat, but they know nothing about meditations and spiritual wisdom; what is called meat, and what is called green vegetables? What leads to sin? After the tradition of the gods, was the rhinoceros slaughtered; And of its flesh burnt sacrifice offered" (SGGS, p.1089). He also condemned rituals, which did not have a scientific basis. In a famous incident, while traveling to the shrine of Lord Jagannath at Puri, he found the ceremonial ritual of the *aarti* as a mechanical exercise. Instead, he said to Mardana, the Muslim bard traveling with him, "The firmament is thy salver; the sun and the moon thy lamps, the galaxy of stars as pearls scattered; the forests thy flowers, but what worship is this, O destroyer of Fear? (SGGS, p.13; translated by Singh, 2004).

Guru Nanak's life and writings thus emphasize that inequalities in society have distinct cultural and ethical dimensions. These dimensions act as drivers for an unsustainable path of development, wherein the disproportionate burden of environmental as well as societal pollution is borne by the most vulnerable and deprived sections of the society. The cultural and social contexts of these varied forms of discrimination need to be understood, and dealt with at the practical level, so that a more socially sustainable model of society may be devised. His teachings enable us to understand that ecological concerns are intrinsically related to larger socioeconomic challenges confronting humanity such as massive poverty, gender discrimination, and widespread socioeconomic inequities. Thus, environmental and societal concerns cannot be treated silos, and a holistic model of sustainable development must bring these two together. Some empirical studies on contemporary ecological challenges have also confirmed to this fact. For instance, Singh (2022a) demonstrates that one of the principal reasons why bioenergy transitions have not been successful in many developing country contexts is since the gendered aspects of use of traditional sources of energy have not been incorporated in the mainstream bioenergy policies. Other studies have shown how indoor air pollution is statistically correlated with household level poverty (Hilton, 2006). However, systematic attempts to integrate the dialectical relationships between socioeconomic and ecological aspects of sustainability remain scare. Guru Nanak's teachings can thus provide guidance to policymakers and organizations dealing with material sustainability issues on how to link ecological challenges with other forms of societal inequities and vulnerabilities in order to create more inclusive and people-centric solutions to contemporary ecological challenges.

2.6 Guru Nanak's Concept of Material Well-Being

In mainstream economics, the only parameter of well-being is regarded as higher-income levels or GDP per capita. However, Guru Nanak developed a more nuanced concept of the complexities and varying conditionalities which underlie human and societal well-being. In his writings he has talked about *dukh* as a fundamental human reality and equated it to "pervasive dissatisfaction" (Kalra et al., 2013). Guru Nanak believed that accumulation of wealth and attachment to material wealth and riches can never lead to human happiness and well-being in the long run. He wrote:

> *Kings, rulers and monarchs enjoy pleasures and gather the poison of Maya.In love with it, they collect more and more, stealing the wealth of others. But even as they look on, Maya cheats them, and they come to regret and repent* (SGGS: 1245).

Guru Nanak believed that the roots of dissatisfaction in society are embedded in *haumai* (selfish self-interest). This understanding of pervasive dissatisfaction in human society owing to self-interest has profound consequences for modern-day concept of human well-being and satisfaction. In modern capitalist economies, "the desire to satisfy systemic, constantly regenerated wants is a major motive of our lives and permeates almost every aspect of our life" (Daniels, 2007). This causes to be enslaved by a vicious cycle of consumption of higher and higher quantities of material goods and services. The cycle of ever-increasing human wants and desires has led to excessive extraction from our natural resource base. The material and energy flows needed to satisfy the needs of the modern consumerist economies have led to exhaustion of many of our key natural resources. In this context, Guru Nanak condemned *moh* (materialism) and *lobh* (human greed) as primal sins. "Pollution of the mind is greed; the pollution of the eyes is to look with covetousness upon another's wealth" (SGGS: 472).

Guru Nanak's philosophy however rejected the notion of "karmic fatalism" or the notion that human suffering and tribulations are a consequence of our actions and behavior in previous births (Singh, Singh, 2022a). Nanak said that right actions and conduct would translate into salvation and attainment of human well-being.

> *As a team of oxen are we driven.*
> *By the ploughman, our teacher.*
> *By the furrows made are thus writ.*
> *Our actions-on earth, our paper.*
> *The sweat of labour is as beads* (SGGS: 144, translated by Singh, 2004)

Guru Nanak proclaimed that the righteous actions would lead us toward a path of *sahaj* (detached balance) between life's desires and complete renunciation (Tatla, 2008). Nanak's life was spent on long journeys (*udasis)* to spread his message far and wide (The Tribune, 2019). Yet, he believed in the ideas of worldliness and giving back to one's community. He rejected asceticism and wrote, "Asceticism lies in remaining pure amidst impurities" (SGGS: 223). In the last part of his life, Nanak established a model society in Kartarpur, which effectively integrated the social and spiritual dimensions of life.

Guru Nanak's teachings on righteousness of human actions, curtailing conspicuous consumption and giving back to society, have profound implications for the current debates surrounding sustainable consumption and production and challenge the mainstream economic maxim which equates material accumulation to satisfaction and improved quality of life (Sirgy, 1998). In contrast, Guru Nanak's philosophy analyze how subjective human well-being is associated with a diverse set of psychological and physical conditions and is closely intertwined in an individual's relationship with his community and the natural environment. The foresightedness of Guru Nanak's is mirrored in several empirical studies which demonstrate how traditional yardsticks of economic growth fail to capture subjective well-being and sustained welfare satisfaction (Frugoli et al., 2015; Di Fabio, 2017). On the other hand, "treadmill of production" (Daniels, 2007) in modern capitalist societies to meet higher and higher level of material acquisition is putting increased pressure on environmental resources eroding our natural resource base. Guru Nanak's philosophy enables to understand that unsustainable patterns of consumption are rooted in a lack of knowledge about the pathway to achieve contentment and satisfaction. His philosophy thus provides guidance to sustainability practitioners and organizations to create awareness about the lacunae of higher and higher levels of material acquisition and instead their target individual preferences to achieve long-term welfare goals.

2.7 Guru Nanak's Conception of Sustainable Societies

At the end of his *udasis*, around 1521, he settled as a farmer in a new town, Kartarpur, on the banks of river Ravi and established it as an eco-sophical community. The Kartarpur model represented a framework of community ownership, voluntary service, and pooling of resources (The Tribune, 2019). It was based on three main practices: *sangat*, *pangat*, and *langar*. Guru Nanak said: "They alone who live by the fruit of their own labour and share its fruit with others have found the right path" (SGGS: 1245). His followers gathered here, "not as a monasthetic order, but as individuals with different capacities working in a co-operative manner" (Tatla, 2008). People from all walks of life participated in the social and spiritual life of Kartarpur, defying the artificial division of caste, class, and gender (Takhar, 2011; Singh, 2022a).

The practice of langar became the defining feature of the Kartarpur community and epitomized the spirit of solidarity, fraternity, and equality (The Tribune, 2019). This system also posed a direct challenge to the institutional practice of untouchability, wherein food cooked by the so-called low-caste populations was deemed unworthy of being consumed by higher-caste communities. In a revolutionary stance, Nanak abolished the rigid caste hierarchy in preparation of food in the community kitchen. Moreover, Guru Nanak's followers would eat together in a *pangat* (row of people) irrespective of their caste status. This practice is followed by Sikh

Gurdwaras even today. As a tribute to Guru Nanak, the practice of langar has been adopted by the Barcelona World Parliament of Religions held every year.

"Guru Nanak's Kartarpur model mirrors the vision of 'needs based societies' where in good are produced only to fulfil basic societal needs, not to achieve the objectives of wealth accumulation and surplus value creation. The Kartarpur model provides us with an exemplar of an inclusive, community-based society, which can help to achieve egalitarianism along with an environmentally friendly development trajectory" (Singh, 2022a, p. 41). In such a "needs-based community" waste of energy, raw materials, and superfluous goods can be brought down considerably, resulting in reduced environmental pressures. Guru Nanak's Kartarpur model thus illustrates how involvement of local communities can help in creating a common social consensus for ecologically sensitive projects through involvement of ground-level stakeholders to develop inclusive and sustainable communities. In recent years, several Sikh organizations are drawing inspiration from these basic principles put forth by Guru Nanak to create new forms of environmental activism (Singh, 2022a). For instance, many Sikh temples have started green *langars* through preparation of organic food in the community kitchen and recycling their waste products (Singh, 2022a). Pingalwara, a charitable Sikh organization, has initiated a new form of community-based program wherein they give training to farmers in natural farming techniques (Mooney, 2018). Another diaspora Sikh organization has developed a social enterprise called *KarmaGrow* which uses the Sikh concept of seva (voluntary service) to promote sustainable farming practices (Mooney, 2018). Sant Seechewal who was awarded as a "Hero of the Environment" by the Time Magazine in 2008 employed the Sikh concept of *sevain* order to mobilize volunteers from 21 surrounding villages to manually clean the Kali Bien River, where Guru Nanak attained spiritual enlightenment (Singh, 2022a).

2.8 Conclusion

The principal objective of this paper was to illustrate how Guru Nanak's philosophical paradigm can charter a new paradigm of sustainable development. The paper identified that Nanak's writings provide us with important insights into the environment-economy-society dynamic. It can help to strengthen our existing understanding of the linkages between natural environment and human society, the conception of material well-being, and the varied forms of societal and environmental pollution which have distinct sociocultural sub-texts. Nanak's worldview not only helps us to identify the existing drivers of unsustainability in society; it also provides a model of a new community-driven "needs"-based society. Nanak's teachings show how collective, social strategic action can modify the social fabric of human society and foster long-term environmental and social sustainability. Guru Nanak's understanding of the dimensions of material well-being in society and the cultural aspects of social inequities in given societies are especially pertinent in present times.

Guru Nanak's philosophical paradigm has profound implications for organizations and national governments perusing a sustainable development agenda. First, in contrast to mainstream economics frameworks which treat protection of environment and economic growth as "tradeoffs" (Baumol & Oates, 1988; Hoffman, 2011), Sikh teachings and principles focus on human well-being and protection of nature as "complementary" to each other. These Sikh traditions thus provide inspiration to organizations to embed sustainability as an essential component of their organizational goals. This is illustrated in a limited way through the exemplars of Sikh organizations and individuals pursuing environmentally friendly projects, presented in the section above. Secondly, Guru Nanak's philosophy provides an understanding of the linkages between socioeconomic vulnerabilities and environmental catastrophes. It thus highlights how people-centric frameworks for sustainable development need to take account of the disproportionate impact of environmental challenges on the economically and socially vulnerable groups in society. However, mainstream sustainability policies continue to be framed within a "one size fits all framework," without taking account of the regional and socioeconomic constraints particular to a given region, leading to limited social acceptability of these projects. Thirdly, Guru Nanak's Kartarpur model provides a vision of a needs-based society and is being used by Sikh organizations to create collective praxis for environmentally friendly projects such as organic farming, renewable energy deployment, recycling of waste, and tree plantations. This model can provide inspiration to non-Sikh organizations as well, in creation of sustainable communities as well as building a social consensus for pro-environmental projects.

Some critics argue that Guru Nanak's vision can only provide a utopian vision of society, which is not a realistic alternative economic model. This is especially true because Guru Nanak's vision of sustainable egalitarian society has not been followed in praxis by the Sikh community at large. However, I believe that the superimposition of Guru Nanak's wisdom on comprehensive sustainability frameworks like the three pillars model will help us to formulate specific strategic actions and create a new blueprint for sustainable economic development in the long run.

References

Adman, F., & Ozkaynak, B. (2002). The economics-environment relationship: Neoclassical, institutional, and Marxist approaches. *Studies on Political Economy, 69*, 109–135.

ARC. (n.d.-a). *Alliance of religion and conservation-history*. http://www.arcworld.org/about3a59.html?pageID=2

ARC. (n.d.-b). *Sikh faith statement*. http://www.arcworld.org/faiths462d.html?pageID=73

Ashton, G. (2014). Hinduism and environmental ethics: Law, literature, and philosophy by Christopher G. Framarin. *Environmental Ethics, 36*(3), 369–372.

Barnosky, A. D., & Hadly, E. A. (2016). *Tipping point for planet Earth: How close are we to the edge?* Macmillan.

Baumol, W. J., & Oates, W. E. (1988). *The theory of environmental policy* (2nd ed.). Cambridge University Press.

Castro, C. J. (2004). Sustainable development: Mainstream and critical perspectives. *Organization & Environment, 17*(2), 195–225.
Cropper, M. L., & Oates, W. E. (1992). Environmental economics: A survey. *Journal of Economic Literature, 30*(2), 675–740.
Daniels, P. L. (2007). Buddhism and the transformation to sustainable economies. *Society and Economy, 29*(2), 155–180.
Daniels, P. L. (2010). Climate change, economics and Buddhism-Part I: An integrated environmental analysis framework. *Ecological Economics, 69*(2010), 952–961.
Daniels, P. (2011). Buddhism and sustainable consumption. In *Ethical principles and economic transformation-a Buddhist approach* (pp. 35–60). Springer.
Dawson, A. (2013). Biohazard: The catastrophic temporality of green capitalism. *Social Text, 31*(1114), 53.
Di Fabio, A. (2017). The psychology of sustainability and sustainable development for well-being in organizations. *Frontiers in Psychology, 8*, 1534.
EPGN. (n.d). *History of the EGPN*. https://www.greenpilgrimageeurope.net/egpn-history.html
Frugoli, P. A., Almeida, C. M. V. B., Agostinho, F., Giannetti, B. F., & Huisingh, D. (2015). Can measures of well-being and progress help societies to achieve sustainable development? *Journal of Cleaner Production, 90*, 370–380.
Hanley, N., Shogren, J. F., & White, B. (1997). *Environmental economics in theory and practice* (pp. 179–180). Macmillan.
Helfaya, A., Kotb, A., & Hanafi, R. (2018). Qur'anic ethics for environmental responsibility: Implications for business practice. *Journal of Business Ethics, 150*(4), 1105–1128.
Henisz, W. J. (2000). The institutional environment for economic growth. *Economics and Politics, 12*(1), 1–31.
Hilton, F. G. (2006). Poverty and pollution abatement: Evidence from lead phase-out. *Ecological Economics, 56*(1), 125–131.
Hoffman, A. J. (2011). Talking past each other? Cultural framing of skeptical and convinced logics in the climate change debate. *Organization & Environment, 24*(1), 3–33.
Jacobs, M. (2012). *Green growth: Economic theory and political discourse* (Working Paper No. 108). Centre for Climate Change Economics and Policy, Grantham Research Institute on Climate Change and the Environment.
Janicke, M. (2012). Green growth: From a growing eco-industry to economic sustainability. *Energy Policy, 48*, 13–21.
Jenkins, W. (2013). *Ecologies of grace: Environmental ethics and Christian theology*. Oxford University Press.
Johnston, L. F. (2014). *Religion and sustainability: Social movements and the politics of the environment*. Routledge.
Kalra, G., Bhui, K., & Bhugra, D. (2013). Does Guru Granth Sahib describe depression? *Indian Journal of Psychiatry, 55*(6), 195–200.
Khaleel, F., & Shobha, K. (2021). Ontological authority of sustainability in Hindu traditions: The art of planetary maintenance. In *Foundations of a sustainable economy* (pp. 96–104). Routledge.
Lamentowicz, M., Kołaczek, P., Mauquoy, D., Kittel, P., Łokas, E., Słowiński, M., et al. (2019). Always on the tipping point–A search for signals of past societies and related peatland ecosystem critical transitions during the last 6500 years in N Poland. *Quaternary Science Reviews, 225*, 105954.
Lenton, T. M., et al. (2008). Tipping elements in the Earth's climate system. *Proceedings of the National Academy of Science (PNAS), 105*(6), 1786–1793. (Available online at http://www.pnas.org/content/105/6/1786.full.pdf) [Accessed on 6/12/2015]
Lenton, T. M., Rockström, J., Gaffney, O., Rahmstorf, S., Richardson, K., Steffen, W., & Schellnhuber, H. J. (2019). Climate tipping points—Too risky to bet against. *Nature, 575*, 592.
Living Planet Report. (2022). *Living planet report, 2022*. WWF International.

Lourdunathan, S. (2012). Ecosophical concerns in the Sikh tradition. In C. C. Cain (Ed.), *Many heavens: One earth: Readings on religion and environment*. Lexington Books.
Mooney, N. (2018). Sikh millennials engaging the earth: Sikhi, environmental activism, and eco-enchantment. *Sikh Formations, 14*(3–4), 315–338.
Neven, D., Krzysztof, U., & Huisingh, D. (2015). Components and structures of the pillars of sustainability. *Journal of Cleaner Production, 88*, 1–12.
Prill, S. E. (2015). Sikhi and sustainability: Sikh approaches to environmental advocacy. *Sikh Formations, 11*(1–2), 223–242.
Purvis, B., Mao, Y., & Robinson, D. (2019). Three pillars of sustainability: In search of conceptual origins. *Sustainability Science, 14*(3), 681–695.
Singh, K. (2004). *A history of the Sikhs: Volume 1: 1469–1839* (2nd ed.). Oxford India Paperbacks.
Singh, N. (2022a). Sikhism and sustainability: New approaches to environmental ethics. In *Faith traditions and practices in the workplace volume I*. Palgrave Macmillan.
Singh, N. (2022b). Bioenergy and ecosocialism. In *The Routledge handbook on ecosocialism* (pp. 320–327). Routledge.
Singh, N., & Singh, P. (2019). Political economy of bioenergy transitions in developing countries: A case study of Punjab, India. *World Development, 124*, 106430.
Sirgy, M. J. (1998). Materialism and quality of life. *Social Indicators Research, 43*(3), 227–260.
Solow, R. (1993). Sustainability: An economist's perspective. In R. Dorfman & N. S. Dorfman (Eds.), *Economics of the environment* (pp. 179–187). Norton.
Takhar, O. K. (2011). Egalitarian hermeneutics from the Bani of Guru Nanak: His attitudes towards caste and females. *Understanding Sikhism, 13*, 42–88.
Tatla, D. S. (2008). Sikhism and development: A review. In *Religions and development, working paper no. 21*. DFID.
The Tribune. (2019). Revolutionary thinker and humanitarian. *The Tribune*, November 10, 2019. https://www.tribuneindia.com/news/archive/features/revolutionary-thinker-humanitarian-858135
Tiba, S., & Omri, A. (2017). Literature survey on the relationships between energy, environment and economic growth. *Renewable and Sustainable Energy Reviews, 69*, 1129–1146.
Tinnell, J. C. (2012). Transversalising the ecological turn: Four components of Félix Guattari's ecosophical perspective. *Deleuze Studies, 6*(3), 357–388.
Virk, H. S. (2018). *Scientific vision of Guru Nanak about the universe, and Stefan Hawking*. Paper presented at the SGGS Khalsa College, New Delhi, October 24, 2018.
World Meteorological Association. (2022). *Provisional State of Global Climate in 2022*. https://public.wmo.int/en/our-mandate/climate/wmo-statement-state-of-global-climate

Part III
Christianity

Chapter 3
Family Firms: The Impact of a Christian Perspective on Environmental Protection and Sustainability Practices

Allan Discua Cruz and M. Isabella Cavalcanti Junqueira

Abstract We focus on sustainability and a Christian perspective in family firms, the world's most common form of business. Based on a review of the literature, we argue that a Christian perspective is a key factor in our understanding of sustainability and business operations. We contend that sustainability in business is impacted by the personal values of family firm owners and their employees. Scholars have found evidence of the relevance of religious faith in decision-making processes of family firm owners. Thus, we explore and theorize how Christian principles influence the sustainability efforts of family firms. We offer a model that attends to our main theoretical question and interrogates how sustainability, influenced by a Christian perspective, is expressed in the context of family firms.

Keywords Sustainability · Family business · Family firm · Christian · Stewardship

3.1 Introduction

While there is no single definition of sustainability, it broadly relates to "enduring into the long term future … [and] to systems and processes that are able to operate and persist on their own over long periods of time" (Robertson, 2017, p. 3). Management scholars agree that sustainability refers to how firms balance environmental integrity, social equity, and economic prosperity to secure intergenerational equity (Bansal & DesJardine, 2014). Compound macro- and microenvironmental

A. Discua Cruz (✉)
Centre for Family Business, Lancaster University Management School, Lancaster, UK
e-mail: a.discuacruz@lancaster.ac.uk

M. I. Cavalcanti Junqueira
LeTourneu University, Longview, TX, USA

Montreat College, Montreat, NC, USA
e-mail: isabellajunqueira@letu.edu

N. Singh et al. (eds.), *Faith Traditions and Sustainability*, Management, Change, Strategy and Positive Leadership, https://doi.org/10.1007/978-3-031-41245-5_3

fluctuations affecting firms have shaped accelerated economic, social, technological, and political changes. As a result, recent studies show that current socioecological and economic challenges are multifaceted and existential threats to society at large (Gunton et al., 2022).

Shifting environmental factors have garnered significant firm challenges that require firm management and adaptation to assure success. While the management literature has begun to investigate the nature of these challenges (Adger et al., 2005; Maniora, 2018), further research is needed to address how firms should respond to factors like environmental degradation and climate change and how shifts in organizational culture can be part of this process (Tyler et al., 2020). Scholars have also begun investigating the underlying mechanisms, including those that draw from a religious faith (Hoffman & Sandelands, 2005), which allow organizations to address environmental concerns under conditions of uncertainty (Muller et al., 2022).

An integral part of everyday operations is attending to trends in the external environment (Wilkinson et al., 2001). To broaden our knowledge of how these challenges are being addressed, recent studies have called for further exploration of family firms – the most common form of business in the world today that intertwines family and firm objectives over time (Howorth & Robinson, 2020). Family firms are ubiquitous and heterogeneous (Memili & Dibrell, 2019), existing and surviving in diverse contexts and guises. They are also the heartbeat of the global economy, ranging from small corner shops in local towns and cities (Cortez Arias & Discua Cruz, 2018) to multinational business conglomerates and groups (Rautiainen et al., 2019).

Since family firms participate in all industry sectors (Baù et al., 2021; Sharma & Sharma, 2021), they have had – and continue to have – a key role in the diverse concerns of sustainability. Their embeddedness, strong visibility, and reputation are not only critical levers as they influence their local communities (Shainaz & Gomez-Mejia, 2021) but also key factors in our understanding of their complex relationship with the environment.

Family firms operate in multiple industries and provide a broad array of products. For instance, Walmart offers products ranging from groceries to car maintenance and everyday fashion to technology. These products can create substantial amounts of pollution and industrial waste through manufacturing processes, transportation of raw goods, packaging materials, and distribution channels (Soppelsa et al., 2021). To enhance efficiency and productivity, some family firms might introduce harmful chemicals and processes to the environment (Cousins et al., 2019). For instance, in the fashion industry, industrial waste can be toxic to the natural world, as well as human life, and production and associated supply chains account for 10% of the global carbon emissions (Brewer, 2019). An inevitable yet unfortunate corollary is that little is known about how family firms might successfully address these environmental concerns.

While diverse theoretical perspectives have been used to elucidate the emergence, development, and survival of family firms (Discua Cruz & Basco, 2018), it is only recently that studies have revealed how and why religious beliefs influence this form of business (Astrachan et al., 2020; Cater III & Alderson, 2022; Sorenson

& Milbrandt, 2022). While recent studies call for a better understanding of the underlying rationales motivating religious individuals to build their ventures (Litz, 2013; Barbera et al., 2020; Rietveld & Hoogendoorn, 2021) and ensure their survival over time (Burton et al., 2022), there is a paucity of research addressing how family firms – upholding a particular religious perspective – address sustainability. In this chapter, we contribute to our understanding of how Christian family firms approach sustainability.

We contend that, in the context of family firms grounded in a Christian perspective, sustainability may entail rethinking business operations (Ewest, 2018). These firms have endured in diverse societies and can be found in diverse industries, including food retailers and technical services (Carradus et al., 2020; Cater III & Alderson, 2022; Discua Cruz, 2013; Wong et al., 2018). While people may be surprised to learn that some of the largest corporations are family-owned firms, such as IKEA, Walmart, and Samsung, other family firms may be well – or less well – known for their long-standing Christian principles and heritage (Cathy, 2007; Higginson & Robertshaw, 2017; Malloch, 2008; Redmer, 2008). Christian principles not only inform assumptions about how systems work (e.g., Chick-fil-A; see Robinson, 2019) but also affect the day-to-day decision-making processes of business leaders, such as how to manufacture and market healthy foods. The central premise of a Christian sustainability perspective is that leaders choose to serve God first and foremost. This perspective can shape distinctive approaches to sustainability (Cafferky, 2012; Wong et al., 2018) while permeating organizational culture with ethical, faith-based belief systems (Weber, 1905/1958, foreword by Tawney). For instance, leaders of Christian family firms may support their communities (Hayes & Robinson, 2011) through charity, problem-solving interventions, and increased awareness of place (Seifert & Shaw, 2013).

We propose a model that addresses our main theoretical question: How is sustainability, influenced by a Christian perspective, expressed in the context of family firms? Our primary focus is thus an understanding of how underlying Christian principles inform family firms as they engage in sustainability efforts in rapidly changing environments.

3.2 Christian Ethics and Sustainability

Extant research has identified a relationship between Christianity and sustainability, suggesting that theoretical perspectives can be useful in revealing the impact of Christian perspectives (Boyd, 1999; Cafferky, 2015; Cundill, 2021; Spence & Brown, 2018). Focusing on Christianity as a relevant influence within sustainability efforts is essential in advancing our understanding of Christian principles and how they may be related to decision-making processes of the owner-managers of family firms (Ibrahim et al., 2008; Werner, 2008). For these leaders, a Christian perspective may lead them to choose sustainability practices that they know family firm members are willing to uphold over time (Muller et al., 2022).

However, a Christian perspective has also been criticized for being an ideology that ultimately leads to environmental degradation (Eckberg & Blocker, 1989; White, 1967). For instance, the history of environmental exploitation in the West is linked with some Christian institutions. Such assumptions of the Christian faith, including a lack of interest in environmental concerns, emerged in the 1970s and 1980s (Hitzhusen, 2007). As a result, scholars have recently called for further research acknowledging and addressing one of the most significant challenges in the study of sustainability and family firms – the assumption that sustainability decisions by leaders of Christian family firms are nonetheless religion-neutral (Discua Cruz, 2018, 2020).

3.3 Christianity and Imago Dei

Christianity is considered one of the major world religions (Woodhead, 2004), broadly conceptualized as the good news concerning the historical figure of Jesus Christ of Nazareth (Gooding & Lennox, 2014). Jesus Christ is understood as engendering that good news: his life (i.e., what he did, taught, and claimed), his death (i.e., what it accomplished), and his resurrection (i.e., what it demonstrated: that his claims were true). At the heart of the Christian faith are two historical facts: as a human, Jesus Christ is understood as being born of the royal lineage of David, and, as the Son of God, he is known for his miracles, his crucifixion, and his ultimate resurrection from the dead (Rom 1:1–4; also see Gooding & Lennox, 2014, p. 2). According to Smart (1971), the focal point of Christianity is the resurrection, good news that justifies Jesus Christ's authority – during and after his ministry – and that releases a force within his followers to not only promulgate the Christian faith but also permeate all aspects of life. Thirty years following his crucifixion, the Gospels were written to highlight the teachings of Jesus Christ, thus establishing the roots of the Christian faith.

The basic Christian principle that directly relates to sustainability is the existence of God revealed by the Scripture (Nash, 1992, p. 52). The ultimate Christian authority is thus the Bible, and any decision or action related to sustainability will be impacted by insights drawn from the Scripture (Nash, 1992). While the Bible has much to impart concerning sustainability, it is often taken for granted or overlooked. One of the most notable features of the teachings of Jesus Christ is the parables, concrete illustrations of messages to his followers (i.e., concerning ethical, societal, and business matters), as well as the nature of the relationship between man and God, the relationship between man and other human beings, and the kingdom of God (Smart, 1971, p. 407).

In the centuries that followed, the Christian vernacular, values, and beliefs permeated the culture and the business activities in regions where it was disseminated (Woodhead, 2004). In the Western world, the influence of Christianity in business is apparent since Christianity does not deny the importance of business nor shy away from enterprise (Rossouw, 1994). However, research investigating the influence of

Christian principles and beliefs at the individual and firm level of analysis has found mixed results: in some countries, its influence is part of the fabric of legal practices (i.e., USA), while in others, including Northern Europe and Asia, influence is often underplayed or rejected (Jiang et al., 2015; Koiranen, 2002).

A Christian perspective of the environment and sustainability first appears in Genesis 1, where one of the most salient arguments can be found:

> So God created man in His own image; in the image of God He created him; male and female He created them. Then God blessed them, and God said to them, "Be fruitful and multiply; fill the earth and subdue it; have dominion over the fish of the sea, over the birds of the air, and over every living thing that moves on the earth." (Gen 1:27–28).

This Christian perspective accounts for human dominion over God's creation; however, it also calls for human care for the environment (Cafferky, 2012). Scholars have contended that the words "subdue" and "dominion" do not grant permission to deplete any part of God's creation (Cafferky, 2012; Hitzhusen, 2007). While subduing the earth involves harnessing the power of various resources on earth through advances in science and technology, dominion does not presume a right to abuse or misuse these resources (Hitzhusen, 2007; White, 1967). God's creation is meant for human use, and defacing, abusing, and destroying the environment is thus considered an act against God. Since air, water, land, plants, and animals are all God's creation and thus good (Gen 1:4–31), humans are called to sustain and preserve all – living in peace, working toward justice, and respecting the natural environment to ensure that it is renewed and sustained. Dominion is not interpreted as the authority to deplete God's creation but rather as the mandate to develop wisdom and the ability to care for it. In the Scripture, various verses explicate the notion of dominion: Ps 24:1–2, Exod 19:5–6, Lev 25:23, Ps 50:9–12, Hag 2:8, Jas 3:7, and Deut 11:10–15. Thus, human decisions to destroy or provide care for all of God's creation, from ecosystems to the climate, are relevant and essential themes for scholarly debate.

In addition to human dominion over God's creation, another fundamental principle of a Christian perspective concerns God revealed in Creation. From the outset of the Scriptures, God is presented as the sovereign creator of all things, with Gen 1:26–27 stating that God created man in his own image (i.e., from the Latin term *imago Dei*). Grudem (1994) contended that *imago Dei* could be interpreted as similar yet not identical: "let us make man to be like us and to represent us" (p. 568). While designing his image-bearers, God also stipulated the responsibilities to fulfill the given charge to us as his image-bearers, the "actions that were morally good in the light of our telos (i.e., purpose) as creatures in Christ" (Lett, 2019). Since humanity represents God when working with and cultivating God's creation for the benefit of the whole of creation, God's purposeful plan also requires focused generosity (Liebengood, 2015), stewardship, and responsible use of resources and capabilities (Bradley, 2020).

The principle of responsibility and the precept of accountability are often used interchangeably. However, the Biblical practice of accountability offers an additional and valuable layer of meaning because it demonstrates that Christians must *give an account* of their decisions and actions (Gates & Steane, 2008), first to God and then to man. Responsibility and accountability thus underpin transparency, which is often associated with ethical practices.

3.4 Christianity and Sustainability

Early Christian literature highlighted that God created the world ex nihilo [from nothing], a critical metaphysical tenet of the Christian worldview: the world is not a purposeless machine but rather a God-ordained order (Nash, 1992). As a result, Christians have an inherent, natural relationship with the environment because God loves and sustains his creation, and Christians also understand creation as having inherent, divine value because it is God's handiwork. Thus, the relationship that Christians have with God's creation must also include actions of care to maintain its natural integrity. Cafferky (2012, p. 145) underscored that sin has not only distorted our relationship with each other but also with the natural environment, leading to ecological imbalances stemming from greed-fueled actions. Resources are thus consumed for economic gain, leaving the earth weakened and challenged as it seeks to heal itself. Following the biblical text of Rom 8:18–27, readers are informed about what has occurred to the environment and what ramifications this has for the future.

Hoffman and Sandelands (2005) also argued that, based on the parables of Jesus Christ (e.g., "Sermon on the Mount"; see Matt 6:26–6:30), aspects related to the environment could act as a model for how to live within God's plan. Throughout the sermons and parables of Jesus Christ, metaphors and images of nature (e.g., birds, flowers, mustard seeds, sheep, shepherds, fish, fishermen, soil, planting, harvesting, grain, wheat, trees, and springs of water) convey moral teachings. Hoffman and Sandelands (2005) argued that this approach supports the Christian worldview whereby God "created man and nature on the same plan, each in relation to the other" (Hoffman & Sandelands, 2005, p. 152).

Scholars argue that Christianity's emphasis on science and technology – a contributor to the current environmental crisis – and the Bible's seeming justification not only remain misunderstood within the literature but also require acknowledgment (Whitney, 2015). The Bible suggests that humans have the freedom to interact with the environment if this freedom is not used as a license to destroy (Cafferky, 2012). One example of how physical spaces can be sustained by diverse actors, including business, is chronicled in the Book of Numbers (i.e., Num 35:2–5). Here, the amount of pasture a town should have to maintain its sustainability is described, showing how enterprises should supply products to a city to support and preserve both its economic well-being and ecological growth (Milgrom, 1982). The remains of such enterprises, found in recent excavations, suggest that family firms influence the sustainable development of community and place (Guijarro, 1997).

Along with exercising dominion over God's creation comes the obligation of stewardship. While humans are challenged to act as trustees of God's creation, using it for their benefit, there is also the expectation that humans are to act responsibly, with the knowledge that they are accountable to God (Wong & Rae, 2011). The obligation of stewardship should thus be also addressed when investigating the relationships between family firms, sustainability, and a Christian perspective.

3.5 Family Firms and Sustainability

Family firms have not gone unnoticed when sustainability and environmental practices have been questioned (Muller et al., 2022). Since family firms of all sizes in various industries are increasingly judged on how they impact the environment, they are also increasingly motivated to implement sustainability practices, such as recycling materials and choosing sustainable suppliers.

Mitigating risk in the natural environment while creating social value over the short and long term (Borga et al., 2009; Slawinski & Bansal, 2015) can be aligned with financial performance goals. Motivations to mitigate such risks stem from both external pressures, such as climate change and industry groups, and internal pressures, such as managerial values and increasing operational efficiencies (Cavalcanti Junqueira, 2021; Muller et al., 2022; Tyler et al., 2020). Family firms choosing to implement sustainable practices may gain competitive advantages by leveraging sustainability for cost advantages or to differentiate from larger competitors (Discua Cruz et al., 2020). For instance, since consumers demand environmentally conscious brands now more than ever, the decision to create sustainable brands and products can lead to higher brand equity (Spielmann et al., 2021).

With limited resources, family firms can struggle in the short term to manage costly operations, equipment, and personnel changes (Campopiano et al., 2020). Short-term costs can thus steer decision-making processes of the leaders of family firms in adverse directions. Changes required for sustainable operations may be viewed as unsound strategic choices because they may also negatively impact value creation over the long term (Williamson et al., 2006). However, rising environmental concerns have opened timely discussions and highlighted practical implications for members of family firms and the communities that they serve (Williams & Schaefer, 2013). As new insights into global challenges are uncovered, family firms closely connected to the natural environment, including those in agricultural (e.g., food production and wineries) and forestry industries (Alsos et al., 2011), are induced to take a closer look and to reflect upon their involvement and motivations to embrace environmental protection and sustainable solutions.

Recent findings suggest that family firms differ from other enterprises in terms of their underlying motivations, objectives, and orientations, which positions these businesses, and their commitment to sustainability, as relevant contexts for study (Westhead & Howorth, 2007). Family firm sustainability practices are often linked to efforts that assure family continuity by nurturing a healthy organization for future generations (Delmas & Gergaud, 2014). These practices may also include committing time and resources (i.e., financial and human) to nurture long-term relationships within the firm and with the broader community (Fitzgerald et al., 2010). Sustainability may thus be viewed as a collective undertaking, with family firms catering to not only the needs of the immediate family but also the concerns of a diverse set of actors and stakeholders (Olson et al., 2003). Early studies have revealed that many family firms rely on dialogue and articulation of their approaches in line with the natural features of their operating places (Roscoe et al., 2013).

Since many family firms are embedded in diverse places and are influenced by natural features, their actions can bring distinctiveness to their localities and surrounding environment (Baù et al., 2021). For instance, rural areas, which tend to harbor family firms, have "location-specific advantages," including natural, social, cultural, and historical characteristics that have developed over time (Cortez Arias & Discua Cruz, 2018). Since these unique features are place-specific, and no place is made up of the same mix of resources, regional advantages can thus offer unique opportunities for family firms to create value and adopt sustainable practices (Korsgaard et al., 2015). As a result, we cannot underestimate the power of family firms to act alongside others to contribute to community sustainability.

Several studies have identified that the long-term orientation of family firms leads to activities that create value not only for the family firms but also for the benefit of their communities, both economically and socially. Since family firms are thus frequently embedded in their communities, it is essential that research addresses the inherent sustainability concerns of the areas in which they operate (Lumpkin & Brigham, 2011; Woodfield et al., 2017). Studies suggest that family firm communities are firmly embedded in the local context, as evidenced by the strong relationships between family startups, community attitudes, and pre-existing small family firms (Bird & Wennberg, 2014). More recently, studies have also suggested that the sustainability practices of family firms may have a stronger relationship to religious beliefs than previously understood (Discua Cruz, 2020). Examples such as the H-E-B Corporation (formerly the Howard E. Butt Grocery Company), merit further exploration. H-E-B is committed to sustainability, claiming responsibility "to take care of the communities where [they] live and work" (H-E-B, 2022). In part, this claim is supported by their continued commitment to providing environmental and conservation grants for multiple endeavors (H-E-B, 2022).

3.6 Family Firms, Sustainability, and Christianity

A fundamental challenge in investigating businesses associated with Christian principles is that there are no official databases of Christian firms (Ibrahim & Angelidis, 2005). However, there is evidence that some family firms do label themselves as "Christian" (Discua Cruz, 2013), although these family firms might demonstrate varying levels of commitment and integration with Christian principles (Discua Cruz, 2013). While this type of research is not straightforward, there is a gap in our understanding of how a Christian perspective can inform the sustainability practices of family firms.

Research has shown that a family firm's strongly upheld values and beliefs can influence sustainability practices (Discua Cruz, 2015, 2018). Such values, which may be closely associated with Christian principles, can influence which products, markets, and business opportunities are evaluated, deemed valuable, and eventually pursued by family firms (Discua Cruz, 2015). Thus, the presence or absence of religious beliefs can influence how family firms approach sustainability (Carradus et al., 2020; Discua Cruz, 2020). To date, the literature has offered some relevant theoretical insights on this topic, striking a balance between fundamental and novel concepts (Hamilton et al., 2017).

While Christian principles can influence firm approaches to sustainability and ethical decision-making, how Christian principles actually influence organizations is a complex process (Weaver & Agle, 2002). For instance, tensions may surface when integrating a Christian perspective with sustainability approaches (Nash, 1994; Nash & McLennan, 2001). In a study of the practices and values of 85 evangelical Christian CEOs, Nash (1994) found several further tensions: "love of God vs. profit," "family and work," "charity and wealth," and "humility and ego." More recently, Nash and McLennan (2001) also revealed that communication voids and conflicts within business and religious communities accentuate these types of tensions, which are associated with different types of leaders and the varying degrees to which they incorporate Christian principles as they manage sustainability issues. While some family firm owners/managers may showcase their interpretation of the Christian faith in their approach to sustainability, others may act as clandestine agents. Further, some non-Christian family firms may also follow Christian principles in their sustainability approach.

Prior studies show that family firm heterogeneity is related to the values of founders or incumbents (Aronoff, 2004; Distelberg & Sorenson, 2009). The founder's values are often palpable and visible in newly created or acquired family firms. Since founders often stay in the family firm for a substantial part of their life, they have a greater chance of shaping and transmitting their values and beliefs to the firm (García-Álvarez & López-Sintas, 2001). Family or non-family successors leading the firm may choose to uphold, complement, or remove instilled Christian principles (Discua Cruz, 2015). There is thus excellent potential for conceptual and empirical studies to explore how Christian principles influence sustainability in family firms.

Family firm founders and incumbents that interweave their Christian faith into their approach to sustainability may engage in diverse forms of integration (Discua Cruz, 2013). Drawing from the conceptual framework of Niebuhr (1951), Discua Cruz (2013) identified five different ways in how Christians family firms balance Christian and cultural values: (1) Christ against family businesses (i.e., Christ is viewed as a challenge, an either-or decision for those involved in family businesses); (2) Christ of family businesses (i.e., Christian principles and values are fundamentally aligned with family business practices); (3) Christ above family business (i.e., Christ is the Christ of family businesses, but He is also above family businesses because His values are above the values of the world); (4) Christ and family businesses in paradox (i.e., Christians leading family businesses are citizens of two worlds that are at odds with each other, which demands that they separate the things of God from the things of the world); and (5) Christ the transformer of the family businesses (i.e., Christians are most concerned with the renewal of family businesses through a process that brings them into alignment with Christian principles). The classifications of Discua Cruz (2013) suggest that leaders of family firms, who profess the Christian faith, may gradually guide their firms to any point along this continuum, which can then influence how they develop the relationship between their family firms and sustainability.

As governance is closely related to sustainability, Solomon (2004) analyzed six family firms from a corporate governance perspective to understand how Christianity influenced legal and environmental compliance, products/services, employee satisfaction, and charitable giving. Solomon (2004) found that the executives leading these firms either followed (1) a low-key approach, in which Christian life and business spheres were separated; (2) a preacher-style approach, in which Christian values were woven into business, with the aim of having non-Christian stakeholders embrace Christianity; or (3) a servant-stewardship approach, in which executives placed stakeholders first in their approach to business. These three approaches incorporate a sense of accountability for others, allowing Christian family firm leaders to rethink the waste their businesses produce. The findings of Solomon (2004) also showed that faith integration could positively influence organizations and be sustained over time.

Non-family Christian leaders may also influence sustainability decisions in non-Christian family firms (Discua Cruz, 2015). Non-Christian family firms that incorporate biblical insights in business activities through managers and/or influential stakeholders may also experience a transformation in the values that guide the organization (Discua Cruz, 2015). Christian principles and values may be unconsciously woven into non-Christian firms through new leaders who are believers in Christ (e.g., Christian successors or appointed leaders). These individuals may then impact how Christian principles are integrated within business activities and which approach to sustainability is taken. Conversely, some family firms founded on Christian principles may eventually be transformed into secular organizations by those who lead or manage the firms and take diverse approaches to sustainability practices. Overall, studies that have addressed sustainability within family firms show that leaders can influence environmental decision-making and practices (Sharma & Sharma, 2021).

A Christian perspective of sustainability in family firms can prompt long-term approaches and multifaceted relationships (Cafferky, 2012, 2015). These distinctions are better understood when we explore contexts where a Christian approach influences sustainability. These contexts are characterized by the wise use of natural resources and the care and preservation of the planet. Christian values can encourage leaders of family firms to both act as stewards of God's creation and demonstrate environmental responsibility.

Recent examples of family firms that have integrated a Christian perspective in their sustainability efforts can be found in several countries worldwide. Zigarelli (2019) found that some family firms, such as Tom's of Maine, Cardone Industries, and Herman Miller, not only link a Christian perspective with sustainability but also frame this relationship within their products – ones that create a better world, protect the planet, and value the care of God's creation. Overall, these family firms not only respect God's creation but also allow profit gains so that they can remain viable over the long term (Ruddell, 2014). These firms use natural resources as efficiently as possible and typically have net-zero goals.

3.7 A Stewardship Perspective

While several theoretical lenses can help frame the integration of religious beliefs in business organizations (Miller & Ewest, 2015), recent studies have highlighted the relevance of mainstream perspectives in family firms – particularly a stewardship perspective (Le Breton-Miller & Miller, 2018) – when elucidating the influence of Christian principles (Wong et al., 2018; Discua Cruz, 2018). Some theoretical perspectives of family firms focus on the misalignment of objectives between owners and employees (Chua et al., 2003), while others highlight the relevance of stewards and stewardship.

In the field of management, stewardship is broadly defined as "the attitudes and behaviors that place the long-term best interests of a group ahead of personal goals that serve an individual's self-interest" (Hernandez, 2008, p. 122). Stewardship theory thus focuses on the actions of individuals who take on a stewardship role in family firms and on the behaviors of owners and employees when objectives and goals are aligned (Davis et al., 2010). Hernandez (2008) contended that stewardship is created through social exchanges between family business owners and employees and often across generations.

Stewardship theory is relevant to sustainability and Christianity since it also has a biblical foundation (Discua Cruz, 2015). Amid contemporary ecological crises, family firms that hold Christian traditions and practices can be appropriately rooted in a historical past and the Scriptures (Discua Cruz, 2015, 2018). For instance, the Bible calls for stewardship of the land and the search for *shalom*, the Hebrew word meaning "peace." According to Wolterstorff and Joldersma (2004), *shalom* also means assembling, aligning, and cultivating an appropriate relationship with God, other humans, nature, and oneself (i.e., in that order). There is also a partaking element in *shalom*: man is an actor and contributor in bringing about *shalom* (Sider, 2020). Beyond the precepts of peace and justice interwoven into the meaning of *shalom*, this word also relates to the practice of responsibility and accountability for the community and the environment. Christians are therefore encouraged to pursue sustainable practices due to the value placed on caring for and preserving God's creation: "The Lord God took the man and put him in the Garden of Eden to *work* it and *take care of* it [emphasis added]" (Gen 2:15).

The Hebrew words for "work" and "take care of" may also be translated as "serve" and "preserve" (Akram & Rashid, 2020; Bradley, 2020; Geneva College, n.d.), thus emphasizing that the word *shalom* is a classification of priorities leveraging potential benefits for communities in which family firms operate (Wolterstorff & Joldersma, 2004). Seeking *shalom*, therefore, implies that family firms strengthen positive relationships with diverse stakeholders while guarding the well-being of the places where they dwell and thrive (Cafferky, 2012).

From a Christian perspective, stewardship goes beyond guardianship, as it should encourage applying helpful business tools, acumen, skills, and expertise. Following a stewardship perspective, Christian principles charge owners of family firms to be diligent and wise investors with what has been entrusted to them, both for their welfare and for those who work with them (Discua Cruz, 2018). The goal is to safeguard existing skills and resources and put them to good use in the organization for the benefit of all. Liang (2011) argues that employees, acting as stewards in family firms, should (1) serve the best interests of owners over the long term; (2) maximize the actual worth of entrusted resources; (3) save strategically for a better, albeit uncertain, future; and (4) invest in relationships and build equity as shrewd counsels and trusted executors of the strategic decisions of the owners. Discua Cruz (2015) highlights that leaders of Christian family firms sometimes support rationales for stewardship by relying on biblical narratives, such as the parable of the talents (Matt 25:14–30) and the portrayal of ideal characters (Genesis 39; 41; 47).

The influence of Christian principles on sustainability within family firms may impact how leaders approach sustainability practices, that is, through the eyes of faith. For instance, when scanning the external environment for opportunities to address environmental concerns, Christian leaders may evaluate opportunities, for both business and family objectives, according to the Christian faith. Christian leaders may also evaluate opportunities beyond straightforward economic perspectives and instead incorporate evaluations of how likely an opportunity will make it more or less difficult to remain faithful to one's beliefs and care for God's creation (Cafferky, 2012).

Family objectives and rationales shaped by a Christian perspective can influence risk-taking behavior and innovation (Discua Cruz, 2015). Risks may increase when economic analyses of market and industry structures do not include value constraints as the basis for developing strategic decisions. As a result, leaders of Christian family firms engaging in certain environmental practices might be perceived as dishonest. Hence, these individuals might try to identify, evaluate, and pursue (or reject) opportunities based on their adherence to Christian principles (Discua Cruz, 2018).

Some leaders of Christian family firms may reject the introduction of products whose production processes or related marketing approaches may work against Christian principles related to sustainability (Paterson et al., 2013). When family members evaluate a new international market, similar rejections may result when practices in the new market are unethical or against founding Christian principles (Wong et al., 2018). Thus, the sustainability of family firms may also be impacted by Christian principles that shape how practices are identified, evaluated, pursued, and introduced.

For some leaders of family firms, Christian principles can constrain available alternatives, making some decisions off-limits when change is needed in areas related to the environment (Cafferky, 2012). Decisions to pursue or not pursue short-term gains can be impacted by long-term obligations to stakeholders, such as employees, customers, suppliers, and society, and how certain decisions could affect the environment. For instance, long-term thinking may include an unwillingness to invest capital in specific industries that are perceived as producing "bad products." A commitment to maintaining moral relationships over the long term (Fitzgerald et al., 2010) can also limit the vision of what companies can become in the future. Against this backdrop, leaders of Christian family firms may evaluate the success of their organizations and the achievement of their strategic goals in terms of their organizational missions. If given the opportunity, leaders may also subordinate organizational goals while amplifying Christian principles that underpin the formation and success of family firms (Cafferky, 2012). Achieving above-average economic returns can be a goal for Christian leaders that can be supported in times of crisis without compromising Christian principles related to the environment (Cathy, 2007; Zigarelli, 2019).

From an economic standpoint, stewardship focuses on "financial assets, physical facilities, products and services, systems, and processes" to benefit several generations (McCuddy & Pirie, 2007, p. 962). Thus, a Christian stewardship perspective calls for those working in family firms to go beyond guarding assets for firm sustainability in times of crisis and extend this view to the immediate environment and related stakeholders. In adverse times, stewardship involves a "balancing of interests," involving a responsibility to God and one's fellow man (Rossouw, 1994). Leaders of Christian family firms might look at these demands from a Christian perspective (Discua Cruz, 2018). A Christian perspective in times of crisis follows the principle of accountability, which is reflected in the level of integration of Christian principles in business (Cafferky, 2012). According to Discua Cruz (2015), biblical teachings underscore the communal and relational nature of accountability, upon which owners and employees of family firms may often rely when addressing sustainability concerns (Deut 20:19; 1 Chr 29:1; Ps 24:1–2; Ps 95:3–5; Exod 23:10–11).

When entrepreneurial leaders integrate Christian principles in their decision-making processes during times of crisis, related traditional yet successful cultural patterns may help family firms overcome diverse challenges, including environmental ones. For instance, H-E-B provides training to its workforce to address the annual season of severe weather in Texas and follows sustainable practices that support the environment and the communities they serve. Since H-E-B's desire to serve is grounded in Christian traditions and the desire for excellence, Christian principles *in business* become a meaningful guide *for business* (Discua Cruz, 2018). Rather than exiting a problematic industry as an easy way out, leaders of Christian family firms may choose to improve the practices of their industry in support of the environment. As stewards, they may procure the best ecological solutions for the environment in which they operate while exercising careful responsibility for God's creation (Parish, 2022).

3.8 A Christian Perspective of Sustainability in Family Firms

Based on the works of Discua Cruz (2013, 2015, 2018), Cafferky (2012), and Wong and Rae (2011), Fig. 3.1 presents our model of a Christian perspective of stewardship-based environmental protection and sustainability practices in family firms. First, the activities of Christian family firms are supported by natural resources from God's creation. While the Scriptures encourage the responsible use of resources to support the production of services and products for the betterment of humanity, Christian family firms must balance responsible use with environmental care. Second, environmental care should include a broader, transformative mission whereby existing products and processes are made less harmful. For instance, some Christian family firms, such as Parts Life, Inc., already develop eco-effective products that create new materials and processes. The skills and knowledge of leaders and employees are put to work for the benefit of the company and the environment, engaging a cyclical approach as much as possible. Waste is a cost; however, repurposed or eliminated waste can be transformed into a financial return. In this context, leaders in Christian family firms must find solutions to environmental problems, which demands stronger links between practitioners, policymakers, and researchers.

For instance, Herman Miller (2023) has revised their manufacturing processes through a "design for environment," implementing environmentally responsible

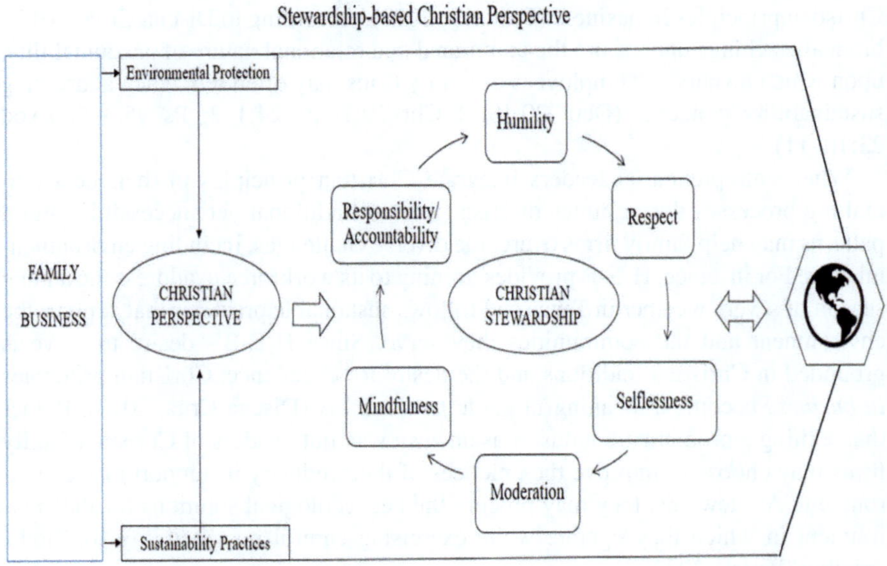

Fig. 3.1 Model of a Christian perspective of stewardship-based environmental protection and sustainability practices in Christian family firms

concepts in the manufacturing cycle, acquiring better materials, and optimizing supply chains. As a result, Herman Miller not only produces products from natural resources that can serve their business but also supports a reclamation process that brings discarded materials back into the product lifecycle with minimal waste and ecological impact. Some byproducts previously discarded or sent to landfills are now repurposed or form the inspiration for new ideas and products, generating new lines of business with an environmental view.

The transformation of existing natural resources must develop alongside an approach for restoration. Consistent with broader biblical themes of the relationship between humans and the environment (e.g., Gen 2:15), our model reveals that the relationship between Christian family firms and the environment is reciprocal. Based on Cafferky (2012, p. 145) and Hoffman and Sandelands (2005), our model also reveals that a Christian-centered view of the environment results in six interrelated dimensions related to the sustainability of Christian family firms: humility, respect, selflessness, moderation, mindfulness, and responsibility. Table 3.1, illustrates our six dimensions of a Christian-centered view of the environment with implications for Christian family firms.

Table 3.1 Dimensions of a Christian-centered view of the environment: implications for Christian family firms

Dimensions	Description	Implications for Christian family firms
Humility	Humility is a central tenet of a Christian perspective. There is an omnipotent and inscrutable God before man and before nature, which highlights significant limitations to man's knowledge and actions. Man's power to harness the environment is, thus, also limited	The multifaceted issues in the natural environment are poorly understood, such as the capacity to provide the resources people need while addressing the pollution humans create. Leaders of Christian family firms should thus take steps to avoid the temptation to discredit the needs of diverse stakeholders who are concerned about the environment or who need help dealing with environmental issues. Lacking knowledge of the impact of human actions compels the practice of precautionary principles and prudence. For instance, Thistle & Twig, a florist shop in Morganton, North Carolina (USA), shows significant consideration for the local rural economy and community. They enjoy working exclusively with local farmers to source seasonal blooms https://www.thistleandtwigflowers.com/

(continued)

Table 3.1 (continued)

Dimensions	Description	Implications for Christian family firms
Respect	Respect is required if we show humility toward each other and the earth, following the expectations of God. The environment has divine value for everyone	Leaders of Christian family firms should produce products and services that remind people that the earth was not created for humans alone but rather for humans and nature to live in a respectful balance. Resources should regard our respect for the environment in the light of how it can be managed as stewards. Products that promote the depletion of resources solely for personal pleasure rather than reflect a stewardship role should be avoided (e.g., forest management). For instance, sidetracked brewery, based in Morganton, North Carolina (USA), aims to better serve its small-town community by providing sustainably sourced, world-class craft beer in a community gathering place https://www.sidetrackedbrew.com/community
Selflessness	Christians should limit desires and make sacrifices if they respect the earth and the God who created it – Especially in cases when the actions of individuals can negatively impact the environment	Leaders of Christian family firms should reduce the impact they have on earth. Every new product or service should be checked for potential threats against nature and/or man, which would constitute a form of disrespect for what God has made. This perspective can be observed in CSR practices, initiatives that encourage care for others, and steward rather than agent organizational behavior (e.g., Cardone Industries; see Redmer, 2008). For instance, interstate batteries, based in Dallas, Texas (USA), serves their communities by honoring a general concern for the environment. They have specific policies on battery safety and responsible recycling practices. They have developed a green standard recycling program that reaches out to organizations and individuals https://www.interstatebatteries.com/recycling
Moderation	Christians should guide others to use only what is needed from the environment. A fundamental teaching of Christianity emphasizes limiting the earth's resources, tempering tendencies for materialism and consumption, and providing checks to economic growth that threatens the environment (Wong & Rae, 2011, p. 233)	By challenging the use of the earth's resources, such as calling for fewer virgin materials and less waste disposal, leaders of Christian family firms should stand in stark contrast to today's society's unrestrained materialism and consumerism. The business philosophies of Christian family firms should reduce the selfish tendency for satisfaction *from nature* and relatively immediate satisfaction *in nature*. This notion would also apply to industries, such as lumber and associated businesses that endeavor to adopt moderation and restoration in their resource approach (e.g., Herman Miller, 2023). For instance, parts life, Inc. extends the life of mission-critical assets through reverse engineering, lifecycle sustainment, acquisition logistics, and obsolescence management https://www.partslifeinc.com/about

(continued)

Table 3.1 (continued)

Dimensions	Description	Implications for Christian family firms
Mindfulness	A Christian perspective calls for thoughtful choices and mindfulness by considering the impact of every action on local communities and society at large	When the environmental impact of actions is known, and when alternative actions are available, leaders of Christian family firms should opt for the latter. A Christian perspective would prompt firms to revisit business policies concerning product packaging and disposable products. For instance, Hobby Lobby, a US-based arts and craft retailer, shrinks its environmental footprint by reducing waste, recycling cardboard, and incorporating reusable boxes and bins. Each year, Hobby Lobby keeps more than 16 thousand tons of cardboard out of landfills https://newsroom.hobbylobby.com/articles/hobby-lobby-recycling-initiative/
Responsibility/accountability	A Christian perspective encourages a mandate of God found in the book of genesis: The restoration of nature is the moral challenge to correct inadvertent damages to the environment. Christians are responsible and accountable for using all available means to recover the balance of nature created by God and reverse any disruptions	Leaders of Christian family firms should voice concerns and offer solutions in industries that harm the environment. Initiatives should prompt engagement in CSR that creates a shared sense of responsibility toward others and the earth (Discua Cruz, 2020). For instance, the Howard E. Butt Grocery company (H-E-B) is the 15th largest grocery chain in the USA and one of the leading chains in Texas. This family firm thrived under Howard E. Butt, a committed Christian and preacher who pioneered mental health support through the H-E-B foundation and Laity Lodge, a spiritual retreat and nature center in the Frio River Canyon (Texas Hill Country River region). Committed to sustainability, H-E-B vows "to take care of the communities where we live and work … [and to] embrace sustainable practices across our business" (H-E-B, 2022). H-E-B also provides several environmental and conservation grants to various endeavors https://www.heb.com/static-page/environment

Note: adapted from Cafferky (2012, p. 145) and Hoffman and Sandelands (2005) Our model reveals that leaders of Christian family firms exercise principles such as moderation and mindfulness when approaching environmental protection and sustainability practices. We define mindfulness in the Christian context as initiating a reflective personal judgment (Vu & Burton, 2020) coupled with community feedback and involvement that can unfurl thoughtful choices by considering the impact of every action on local communities and society at large. We contend that moderation and mindfulness may direct leaders of Christian family firms to approach critical sustainability decisions and actions based on shared knowledge, values, and beliefs of the Christian communities where they cultivate relationships (Dana, 2010; Davidson & Honig, 2003). In this context, gathered knowledge can be used in group actions to serve faith communities and society at large. Future studies are needed to understand this process better

If leaders of Christian family firms recognize the current environmental crisis as a moral, ethical, and faith-based issue, one response may be to engage in specific businesses in some sectors while avoiding others. Another response may be a gradual change in business approaches and processes, targeting wasteful production practices while adhering to and improving respectful and responsible business practices. Since this type of approach does not dismiss secular, widespread movements, such as corporate social responsibility (Matten & Moon, 2008), it is an underlying mechanism that can allow initiatives like CSR to be adopted by Christian family firms. Recent studies have demonstrated that when family firms adopt a stewardship approach to business and are guided by Christian principles, they may develop a unique approach to CSR, which appeals to diverse organizational members aiming to become less self-centered and to care more for others (Discua Cruz, 2020).

3.9 Conclusion

In this chapter, we contend that family firms play a vital role in furthering our understanding of how Christian principles can influence sustainability approaches. Current research often overlooks family firms where the religious beliefs of family members can operate in tandem with business goals. By addressing how and why Christian principles influence sustainability in Christian family firms, we contend that future research will attain a better understanding of sustainability since Christian perspectives and principles can guide decision-making processes and shape sustainable business cultures.

We hope that new theoretical and empirical advances will be inspired by our contributions in this chapter. While some concepts and perspectives we have presented are not new, we believe that future research should continue to investigate the impact of Christian principles as fundamental and foundational aspects that can drive sustainability processes and decision-making. For instance, future investigations should investigate international comparisons in how the Christian faith is expressed in varying contexts around the world. Legal frameworks used in different contexts should be investigated regarding their impact on current assumptions of how Christian principles impact the practices, responsibilities, and accountability of family firms (Gupta & Levenburg, 2010). We also hope our chapter encourages future research that will continue to capture family firms' inherent richness and complexity. Both researchers and practitioners can benefit from expanding their views of family firms to include Christian principles that inform sustainability approaches.

References

Adger, W. N., Arnell, N. W., & Tompkins, E. L. (2005). Successful adaptation to climate change across scales. *Global Environmental Change, 15*(2), 77–86. https://doi.org/10.1016/j.gloenvcha.2004.12.005

Akram, T., & Rashid, S. (2020). *Faith, finance, and economy: Beliefs and economic well-being*. Springer.

Alsos, G. A., Carter, S., & Ljunggren, E. (2011). *The handbook of research on entrepreneurship in agriculture and rural development*. Edward Elgar Publishing.

Aronoff, C. (2004). Self-perpetuation family organization built on values: Necessary condition for long-term family business survival. *Family Business Review, 17*(1), 55–59.

Astrachan, J. H., Binz Astrachan, C., Campopiano, G., & Baù, M. (2020). Values, spirituality and religion: Family business and the roots of sustainable ethical behavior. *Journal of Business Ethics, 163*(4), 637–645. https://doi.org/10.1007/s10551-019-04392-5

Bansal, P., & DesJardine, M. R. (2014). Business sustainability: It is about time. *Strategic Organization, 12*(1), 70–78. https://doi.org/10.1177/1476127013520265

Barbera, F., Shi, H. X., Agarwal, A., & Edwards, M. (2020). The family that prays together stays together: Toward a process model of religious value transmission in family firms. *Journal of Business Ethics, 163*(4), 661–673. https://doi.org/10.1007/s10551-019-04382-7

Baù, M., Block, J., Discua Cruz, A., & Naldi, L. (2021). Bridging locality and internationalization – A research agenda on the sustainable development of family firms. *Entrepreneurship & Regional Development, 33*(7–8), 477–492. https://doi.org/10.1080/08985626.2021.1925846

Bird, M., & Wennberg, K. (2014). Regional influences on the prevalence of family versus non-family start-ups. *Journal of Business Venturing, 29*(3), 421–436. https://doi.org/10.1016/j.jbusvent.2013.06.004

Borga, F., Citterio, A., Noci, G., & Pizzurno, E. (2009). Sustainability report in small enterprises: Case studies in Italian furniture companies. *Business Strategy and the Environment, 18*(3), 162–176. https://doi.org/10.1002/bse.561

Boyd, H. H. (1999). Christianity and the environment in the American public. *Journal for the Scientific Study of Religion, 38*(1), 36–44. https://doi.org/10.2307/1387582

Bradley, A. R. (2020). Biblical stewardship and economic progress. In T. Akram & S. Rashid (Eds.), *Faith, finance, and economy: Beliefs and economic well-being* (pp. 30–51). Springer International Publishing. https://doi.org/10.1007/978-3-030-38784-6

Brewer, M. K. (2019). Slow fashion in a fast fashion world: Promoting sustainability and responsibility. *Laws, 8*(4), 1–9. https://doi.org/10.3390/laws8040024

Burton, N., Vu, M. C., & Cruz, A. D. (2022). Our social legacy will go on: Understanding outcomes of family SME succession through engaged Buddhism. *Journal of Business Research, 143*, 105–118.

Cafferky, M. (2012). *Management: A faith-based perspective*. Pearson Education.

Cafferky, M. E. (2015). Sabbath: The theological roots of sustainable development. *Journal of Biblical Integration in Business, 18*(1), 35–47. https://cbfa-jbib.org/index.php/jbib/article/view/108

Campopiano, G., Calabrò, A., & Basco, R. (2020). The "Most Wanted": The role of family strategic resources and family involvement in CEO succession intention. *Family Business Review, 33*(3), 284–309. https://doi.org/10.1177/0894486520927289

Carradus, A., Zozimo, R., & Discua Cruz, A. (2020). Exploring a faith-led open-systems perspective of stewardship in family businesses. *Journal of Business Ethics, 163*(4), 701–714. https://doi.org/10.1007/s10551-019-04387-2

Cater III, J. J., & Alderson, K. J. (2022). How faith affects the leadership of U.S family firm owners. *Journal of Management, Spirituality & Religion*. 19(4), 390–421. https://doi.org/10.51327/DHPI9374

Cathy, S. T. (2007). *How did you do it, Truett?: A recipe for success*. Looking Glass Books.

Cavalcanti Junqueira, M. I. (2021). Preparation to execution: Orchestrating campaign processes in organization-led crowdfunding. In R. Lenart-Gansiniec & J. Chen (Eds.), *Contributions to finance and Accounting series. Crowdfunding in the public sector* (pp. 43–64). Springer. https://doi.org/10.1007/978-3-030-77841-5_3

Chua, J., Chrisman, J. J., & Steier, L. P. (2003). Extending the theoretical horizons of family business research. *Entrepreneurship, Theory and Practice, 27*(4), 331–338.

Cortez Arias, R. A., & Discua Cruz, A. (2018). Rethinking artisan entrepreneurship in a small island: A tale of two chocolatiers in Roatan, Honduras. *International Journal of Entrepreneurial Behavior & Research, 25*(4), 633–651. https://doi.org/10.1108/IJEBR-02-2018-0111

Cousins, E. M., Richter, L., Cordner, A., Brown, P., & Diallo, S. (2019). Risky business? Manufacturer and retailer action to remove per-and polyfluorinated chemicals from consumer products. *NEW SOLUTIONS: A Journal of Environmental and Occupational Health Policy, 29*(2), 242–265.

Cundill, G. (2021). ESG: An acronym for Christians to embrace? *Faith in Business Quarterly, 21*(3), 16–19.

Dana, L. P. (Ed.). (2010). *Entrepreneurship and religion*. Edward Elgar Publishing.

Davidson, P., & Honig, B. (2003). The role of social and human capital among nascent entrepreneurs. *Journal of Business Venturing, 18*(3), 310–331.

Davis, J. H., Allen, M. R., & Hayes, H. D. (2010). Is blood thicker than water? A study of stewardship perceptions in family business. *Entrepreneurship Theory and Practice, 34*(6), 1093–1116. https://doi.org/10.1111/j.1540-6520.2010.00415.x

Delmas, M. A., & Gergaud, O. (2014). Sustainable certification for future generations: The case of family business. *Family Business Review, 27*(3), 228–243. https://doi.org/10.1177/0894486514538651

Discua Cruz, A. (2013). Christian family businesses: Opportunities for further research. *Journal of Biblical Integration in Business, 16*(2), 7–28.

Discua Cruz, A. (2015). Rethinking family businesses through a Christian perspective. *Faith in Business Quarterly, 17*(1), 23–30.

Discua Cruz, A. (2018). Faith, family and work: A Christian perspective. In T. Ewest (Ed.), *Faith and work: Christian perspectives, research and insights into the movement* (pp. 151–167). Information Age Publishing, Inc.

Discua Cruz, A. (2020). There is no need to shout to be heard! The paradoxical nature of corporate social responsibility (CSR) reporting in a Latin American family small and medium-sized enterprise (SME). *International Small Business Journal, 163*, 701–714. https://doi.org/10.1177/0266242619884852

Discua Cruz, A., & Basco, R. (2018). Family perspective on entrepreneurship. In R. V. Turcan & N. M. Fraser (Eds.), *The Palgrave handbook of multidisciplinary perspectives on entrepreneurship* (pp. 147–175). Palgrave Macmillan. https://doi.org/10.1007/978-3-319-91611-8_8

Discua Cruz, A., Centeno Caffarena, L., & Vega Solano, M. (2020). Being different matters! A closer look into product differentiation in specialty coffee family farms in Central America. *Cross Cultural & Strategic Management, 27*(2), 165–188. https://doi.org/10.1108/CCSM-01-2019-0004

Distelberg, B., & Sorenson, R. L. (2009). Updating systems concepts in family businesses: A focus on values, resource flows, and adaptability. *Family Business Review, 22*(1), 65–81. https://doi.org/10.1177/0894486508329115

Eckberg, D. L., & Blocker, T. J. (1989). Varieties of religious involvement and environmental concerns: Testing the Lynn White thesis. *Journal for the Scientific Study of Religion, 28*(4), 509–517. https://doi.org/10.2307/1386580

Ewest, T. (2018). *Faith and work: Christian perspectives, research and insights into the movement*. Information Age Publishing, Inc.

Fitzgerald, M. A., Haynes, G. W., Schrank, H. L., & Danes, S. M. (2010). Socially responsible processes of small family business owners: Exploratory evidence from the National Family Business Survey. *Journal of Small Business Management, 48*(4), 524–551. https://doi.org/10.1111/j.1540-627X.2010.00307.x

García-Álvarez, E., & López-Sintas, J. (2001). A taxonomy of founders based on values: The root of family business heterogeneity. *Family Business Review, 14*(3), 209–230.

Gates, D., & Steane, P. (2008). Ambiguities of justice in a global marketplace: How are ethical and theological considerations relevant to policymakers? *Management Decision, 46*(8), 1146–1165. https://doi.org/10.1108/00251740810901354

Geneva College. (n.d.). *Christians and the environment*. Geneva College. https://www.geneva.edu/community/environmental-stewardship/why_care

Gooding, D., & Lennox, J. (2014). *The definition of Christianity*. Myrtlefield House.

Grudem, W. A. (1994). *Systematic theology: An introduction to biblical doctrine*. Zondervan.

Guijarro, S. (1997). The family in first-century Galilee. In H. Moxnes (Ed.), *Constructing early Christian families: Family as social reality and metaphor* (pp. 42–65). Taylor & Francis.

Gunton, R. M., Hejnowicz, A. P., Basden, A., van Asperen, E. N., Christie, I., Hanson, D. R., & Hartley, S. E. (2022). Valuing beyond economics: A pluralistic evaluation framework for participatory policymaking. *Ecological Economics, 196*, 107420. https://doi.org/10.1016/j.ecolecon.2022.107420

Gupta, V., & Levenburg, N. (2010). The catholic spirit and family business: Contrasting Latin America, Eastern Europe, and Southern Europe. In A. Stewart, G. T. Lumpkin, & J. A. Katz (Eds.), *Entrepreneurship and family business* (Vol. 12, pp. 185–228). Emerald Group Publishing Limited. https://doi.org/10.1108/S1074-7540(2010)0000012009

Hamilton, E., Discua Cruz, A., & Jack, S. (2017). Re-framing the status of narrative in family business research: Towards an understanding of families in business. *Journal of Family Business Strategy, 8*(1), 3–12. https://doi.org/10.1016/j.jfbs.2016.11.001

Hayes, R. N., & Robinson, J. A. (2011). A research note on institutional logics and entrepreneurial action: The case of black church organizations. *Journal of Developmental Entrepreneurship, 16*(04), 499–515. https://doi.org/10.1142/S1084946711001963

H-E-B. (2022). *Our sustainability journey*. Our Texas Our Future. https://ourtexasourfuture.com/

Herman Miller. (2023). *Sustainability*. https://www.hermanmiller.com/better-world/sustainability/

Hernandez, M. (2008). Promoting stewardship behavior in organizations: A leadership model. *Journal of Business Ethics, 80*(1), 121–128. https://doi.org/10.1007/s10551-007-9440-2

Higginson, R., & Robertshaw, K. (2017). *A voice to be heard. Christian entrepreneurs living out their faith*. InterVarsity Press.

Hitzhusen, G. E. (2007). Judeo-Christian theology and the environment: Moving beyond scepticism to new sources for environmental education in the United States. *Environmental Education Research, 13*(1), 55–74.

Hoffman, A. J., & Sandelands, L. E. (2005). Getting right with nature: Anthropocentrism, ecocentrism, and theocentrism. *Organization & Environment, 18*(2), 141–162. https://doi.org/10.1177/1086026605276197

Howorth, C., & Robinson, N. (2020). *Family business*. Routledge.

Ibrahim, N. A., Howard, D. P., & Angelidis, J. P. (2008). The relationship between religiousness and corporate social responsibility orientation: Are there differences between business managers and students? *Journal of Business Ethics, 78*(1–2), 165–174. https://doi.org/10.1007/s10551-006-9321-0

Ibrahim, N., & Angelidis, J. (2005). The Long-Term Performance of Small Businesses: Are there Differences Between "Christian-Based" Companies and their Secular Counterparts? *Journal of Business Ethics, 58*(1), 187–193. https://doi.org/10.1007/s10551-005-1413-8

Jiang, F., Jiang, Z., Kim, K. A., & Zhang, M. (2015). Family-firm risk-taking: Does religion matter? *Journal of Corporate Finance, 33*, 260–278. https://doi.org/10.1016/j.jcorpfin.2015.01.007

Koiranen, M. (2002). Over 100 years of age but still entrepreneurially active in business: Exploring the values and family characteristics of old Finnish family firms. *Family Business Review, 15*(3), 175–188.

Korsgaard, S., Müller, S., & Tanvig, H. W. (2015). Rural entrepreneurship or entrepreneurship in the rural – Between place and space. *International Journal of Entrepreneurial Behavior & Research, 21*(1), 5–26. https://doi.org/10.1108/IJEBR-11-2013-0205

Le Breton-Miller, I., & Miller, D. (2018). Looking back at and forward from: "Family governance and firm performance: Agency, stewardship, and capabilities". *Family Business Review, 31*(2), 229–237. https://doi.org/10.1177/0894486518773850

Lett, J. (2019). Barth on the ethics of creation. In G. Hunsinger & K. L. Johnson (Eds.), *The Wiley Blackwell companion to Karl Barth* (Barth and dogmatics) (Vol. 1, pp. 369–380). WILEY Blackwell.

Liang, E. P. (2011). The global financial crisis: Biblical perspectives on corporate finance. *Journal of Biblical Integration In Business, 12*, 48–61.

Liebengood, K. D. (2015, October). Paul's expectations of generosity. *Christian Reflection* (Generosity), 19–28. https://www.baylor.edu/content/services/document.php/253627.pdf

Litz, R. A. (2013). Leaving the godfather to follow God the father: Successor generation conversion in a mob family. *Journal of Management, Spirituality & Religion, 10*(2), 183–211. https://doi.org/10.1080/14766086.2012.758053

Lumpkin, G. T., & Brigham, K. H. (2011). Long-term orientation and intertemporal choice in family firms. *Entrepreneurship Theory and Practice, 35*(6), 1149–1169. https://doi.org/10.1111/j.1540-6520.2011.00495.x

Malloch, T. R. (2008). *Spiritual enterprise: Doing virtuous business*. Encounter Books.

Maniora, J. (2018). Mismanagement of sustainability: What business strategy makes the difference? Empirical evidence from the USA. *Journal of Business Ethics, 152*(4), 931–947.

Matten, D., & Moon, J. (2008). "Implicit" and "Explicit" CSR: A conceptual framework for a comparative understanding of corporate social responsibility. *Academy of Management Review, 33*(2), 404–424. https://doi.org/10.5465/amr.2008.31193458

McCuddy, M. K., & Pirie, W. L. (2007). Spirituality, stewardship, and financial decision-making. *Managerial Finance, 33*(12), 957–969. https://doi.org/10.1108/03074350710831738

Memili, E., & Dibrell, C. (Eds.). (2019). *The Palgrave handbook of heterogeneity among family firms*. Palgrave Macmillan. https://doi.org/10.1007/978-3-319-77676-7

Milgrom, J. 1923–2010. (1982). The Levitic town: An exercise in realistic planning. *The Journal of Jewish Studies, 33*(1/2), 185–188. https://www-jjs-online-net.login.ezproxy.library.ualberta.ca/archives/fulltext/1040

Miller, D. W., & Ewest, T. (2015). A new framework for analyzing organizational workplace religion and spirituality. *Journal of Management, Spirituality & Religion, 12*(4), 305–328. https://doi.org/10.1080/14766086.2015.1054864

Muller, C. G., Canale, F., & Discua Cruz, A. (2022). Green innovation in the Latin American agri-food industry: Understanding the influence of family involvement and business practices. *British Food Journal, 124*(7), 2209–2238. https://doi.org/10.1108/BFJ-09-2021-0994

Nash, R. H. (1992). *Worldviews in conflict: Choosing Christianity in the world of ideas*. Zondervan Publishing House.

Nash, L. (1994). *Believers In business*. Thomas Nelson Publishers.

Nash, L., & McLennan, S. (2001). *Church on Sunday, work on Monday: The challenge of fusing Christian values with business life*. Jossey-Bass.

Niebuhr, R. H. (1951). *Christ and culture*. Harper & Row.

Olson, P. D., Zuiker, V. S., Danes, S. M., Stafford, K., Heck, R. K. Z., & Duncan, K. A. (2003). The impact of the family and the business on family business sustainability. *Journal of Business Venturing, 18*(5), 639–666.

Parish, D. (2022). Seeking the welfare of a city. *Faith in Business Quarterly, 21*(4), 23–26.

Paterson, T. A., Specht, D., & Duchon, D. (2013). Exploring costs and consequences of religious expression in family businesses. *Journal of Management, Spirituality & Religion, 10*(2), 138–158. https://doi.org/10.1080/14766086.2012.758051

Rautiainen, M., Rosa, P., Pihkala, T., Parada, M. J., & Discua Cruz, A. (Eds.). (2019). *The family business group phenomenon: Emergence and complexities*. Palgrave Macmillan. https://doi.org/10.1007/978-3-319-98542-8

Redmer, T. A. O. (2008). Case study: Cardone Industries, chaplain program. *Christian Business Academy Review, 3*(1), 62–68.

Rietveld, C. A., & Hoogendoorn, B. (2021). The mediating role of values in the relationship between religion and entrepreneurship. *Small Business Economics, 58*, 1309–1335. https://doi.org/10.1007/s11187-021-00454-z

Robertson, M. (2017). *Sustainability Principles and Practice* (2nd ed.). Routledge. https://doi.org/10.4324/9780203768747

Robinson, S. (2019). *Covert cows and Chick-fil-A: How faith, cows, and chicken built an iconic brand*. Thomas Nelson.

Roscoe, P., Discua Cruz, A., & Howorth, C. (2013). How does an old firm learn new tricks? A material account of entrepreneurial opportunity. *Business History, 55*(1), 53–72. https://doi.org/10.1080/00076791.2012.687540

Rossouw, G. J. (1994). Business ethics: Where have all the Christians gone? *Journal of Business Ethics, 13*(7), 557–570. https://doi.org/10.1007/bf00881301

Ruddell, L. (2014). *Business ethics—Faith that works* (2nd ed.). WestBow Press.

Seifert, J. M., & Shaw, B. R. (2013). Tending our Patch of Creation: Engaging Christians in environmental Stewardship through sense of place. *Journal for the Study of Religion, Nature and Culture, 7*(3), 265–288. https://doi.org/10.1558/jsrnc.v7i3.265

Shainaz, F., & Gomez-Mejia, L. (2021). Can family firms nurture socioemotional wealth in the aftermath of Covid-19? Implications for research and practice. *Business Research Quarterly, 24*(3), 249–257. https://doi.org/10.1177/23409444211008907

Sharma, P., & Sharma, S. (Eds.). (2021). *Pioneering family firms' sustainable development strategies*. Edward Elgar Publishing Ltd.

Sider, R. J. (2020). Christian faith and economics. In T. Akram & E. Rashid (Eds.), *Faith, finance, and economy: Beliefs and economic well-being* (pp. 15–30). Palgrave Macmillan.

Slawinski, N., & Bansal, P. (2015). Short on time: Intertemporal tensions in business sustainability. *Organization Science, 26*(2), 531–549. https://doi.org/10.1287/orsc.2014.0960

Smart, N. (1971). *The religious experience of mankind*. William Collins Sons & Co. Ltd.

Solomon, L. D. (2004). *Evangelical Christian executives: A new model for business corporations*. Transaction Publishers.

Soppelsa, M. E., Lozano-Gracia, N., & Xu, L. C. (2021). The effects of pollution and business environment on firm productivity in Africa. *International Regional Science Review, 44*(2), 203–228.

Sorenson, R. L., & Milbrandt, J. M. (2022). Family social capital in family business: A faith-based values theory. *Journal of Business Ethics*. https://doi.org/10.1007/s10551-022-05110-4

Spence, M. B., & Brown, L. W. (2018). Theology and corporate environmental responsibility: A biblical literalism approach to creation care. *Journal of Biblical Integration in Business, 21*(1), 45–54. https://cbfa-jbib.org/index.php/jbib/article/view/492

Spielmann, N., Discua Cruz, A., Tyler, B. B., & Beukel, K. (2021). Place as a nexus for corporate heritage identity: An international study of family-owned wineries. *Journal of Business Research, 129*, 826–837. https://doi.org/10.1016/j.jbusres.2019.05.024

Tyler, B., Lahneman, B., Beukel, K., Cerrato, D., Minciullo, M., Spielmann, N., & Discua Cruz, A. (2020). SME managers' perceptions of competitive pressure and the adoption of environmental practices in fragmented industries: A multi-country study in the wine industry. *Organization & Environment, 33*(3), 437–463. https://doi.org/10.1177/1086026618803720

Vu, M. C., & Burton, N. (2020). Mindful reflexivity: Unpacking the process of transformative learning in mindfulness and discernment. *Management Learning, 51*(2), 207–226.

Weaver, G. R., & Agle, B. R. (2002). Religiosity and ethical behavior in organizations: A symbolic interactionist perspective. *Academy of Management Review, 27*(1), 77–97. https://doi.org/10.5465/AMR.2002.5922390

Weber, M. (with Tawney, R. H). (1958). Foreword. In: *The Protestant ethic and the spirit of capitalism* (T. Parsons, Trans., pp. X–X. Scribner library). (Original work published 1905).

Werner, A. (2008). The influence of Christian identity on SME owner–managers' conceptualisations of business practice. *Journal of Business Ethics, 82*(2), 449–462.

Westhead, P., & Howorth, C. (2007). "Types" of private family firms: An exploratory conceptual and empirical analysis. *Entrepreneurship & Regional Development, 19*(5), 405–431.

White, L., Jr. (1967). The historical roots of our ecologic crisis. *Science, 155*(3767), 1203–1207. https://doi.org/10.1126/science.155.3767.1203

Whitney, E. (2015). Lynn White Jr.'s 'The historical roots of our ecologic crisis' after 50 years. *History Compass, 13*(8), 396–410. https://doi.org/10.1111/hic3.12254

Wilkinson, A., Hill, M., & Gollan, P. (2001). The sustainability debate. *International Journal of Operations & Production Management, 21*(12), 1492–1502. https://doi.org/10.1108/01443570110410865

Williams, S., & Schaefer, A. (2013). Small and medium-sized enterprises and sustainability: Managers' values and engagement with environmental and climate change issues. *Business Strategy and the Environment, 22*(3), 173–186. https://doi.org/10.1002/bse.1740

Williamson, D., Lynch-Wood, G., & Ramsay, J. (2006). Drivers of environmental behaviour in manufacturing SMEs and the implications for CSR. *Journal of Business Ethics, 67*(3), 317–330. https://doi.org/10.1007/s10551-006-9187-1

Wolterstorff, N., & Joldersma, C. (2004). *Educating for Shalom: Essays on Christian higher education*. William B. Eerdmans Publishing Company.

Wong, K. L., & Rae, S. B. (2011). *Business for the common good: A Christian vision for the marketplace*. InterVarsity Press.

Wong, N. D., Smith, A., & Popp, A. (2018). Religiosity, emotional states, and strategy in the family firm: Edm. Schluter & Co Ltd., 1953–1980. *Entreprises et Histoire, 91*, 98–125. https://doi.org/10.3917/eh.091.0098

Woodfield, P., Woods, C., & Shepherd, D. (2017). Sustainable entrepreneurship: Another avenue for family business scholarship? *Journal of Family Business Management, 7*(1), 122–132. https://doi.org/10.1108/JFBM-12-2015-0040

Woodhead, L. (2004). *An introduction to Christianity*. Cambridge University Press.

Zigarelli, M. (2019). *Christian-owned companies: What does it look like when a follower of Jesus runs a business?* 9 to 5 Media.

Chapter 4
Benedictine Perspective on Sustainability: Economic, Environmental, and Social Inspiration for Organizations

Sheila K. Hanson and Ksenia Keplinger

Abstract Sustainability is an important topic in times of environmental and ecological concerns. However, sustainability existed before it gained traction in the modern business world as there is a demonstrated 1500-year history and lived experience of sustainability in Benedictine organizations that are economically self-sufficient and have existed for centuries. Benedictine organizations follow the Rule of St. Benedict (RSB), an ethical code established in the sixth century. This ethical code is relevant as this document has guided for Benedictine sustainability practices worldwide in various industries. Thus, their sustainability practices are worth considering in how they might generalize to modern secular organizations. This chapter covers the main ethical principles found in the RSB that promote sustainable practices in economic, environmental, and social ways. The Benedictine view of sustainability provides lessons and practical implications to current management and organizational behavior scholars and practitioners.

Keywords Benedictine · Ethics · Sustainability · Sustainable development · The Rule of St. Benedict

4.1 Introduction

Sustainability offers a balanced approach towards business with economic, social, and environmental pillars (Carroll, 2021; Purvis et al., 2018). These pillars represent the triple-bottom line for organizations, a phrase coined by

S. K. Hanson (✉)
University of North Dakota, Grand Forks, ND, USA
e-mail: sheila.hanson@und.edu

K. Keplinger
Max Planck Institute for Intelligent Systems, Stuttgart, Germany
e-mail: kkeplinger@is.mpg.de

N. Singh et al. (eds.), *Faith Traditions and Sustainability*, Management, Change, Strategy and Positive Leadership, https://doi.org/10.1007/978-3-031-41245-5_4

Elkington (1994, 2007). Particularly in the business environment, sustainability is a sound strategy that allows organizations to secure economic value while also reaching social and environmental goals (Case & Chavez, 2017; Laszlo, 2008). Sustainability involves both a present and future orientation. The latter is important because striving toward long-term survival while also focusing on the present earnings and performance is a challenge. Sustainability was initially introduced as a new business concept in the 1950s, gained traction in the 1970s, yet scholars continue to refine the definitions and scope of terms like sustainability and corporate social responsibility (CSR) for business applications today (Carroll, 2021). Within the pillars of environmental, economic, and social sustainability, all are important in the balanced approach to sustainability exhibited by the Benedictines; however, the focus of this chapter weighs more heavily on environmental sustainability, since that is a pressing societal concern and the theme of this book. The modern history of sustainability concepts in business is relatively short compared to the nearly 1500-year history and lived experience of sustainability in long-standing Benedictine organizations. Benedictine organizations appear as part of the economic, environmental, and social landscape worldwide in Asia, Africa, Europe, Central America, North America, South America, and Australia/Oceania (OSB, 2022a). There are approximately 7500 monks or brothers in 400 monasteries as well as 13,000 nuns or sisters (OSB, 2022b). These organizations have survived over multiple centuries despite world wars, changes in political systems, financial crises, plagues, and pandemics. One of the main principles for the Benedictines is working sustainably by securing resources for future generations (Keplinger et al., 2016). Given the Benedictines' lived experience of sustainability in different industries and various cultures over an extended period of time, their sustainability practices are worth investigating and contemplating. In common, the Benedictines follow Christian principles and the Rule of St. Benedict (RSB), an ethical code established in the sixth century. This ethical code is important, since religion, spirituality, and ethics are in the forefront of scholarly discourse due to ongoing challenges with sustainability sometimes attributed to a lack of ethical behavior (Aversano-Dearborn et al., 2013; Inauen et al., 2010; McDaniel, 2002; Orr, 2002).

In this chapter, we describe the main ethical principles transmitted through the RSB that promote sustainable practices and how the Benedictine view of sustainability provides lessons and practical implications to current management and organizational behavior scholars and practitioners. We then investigate how the Benedictines fit within the concept of sustainable development. Then we proceed to outline the economic, environmental, and social cornerstones of sustainability from a Benedictine perspective. We illustrate how the concept of Benedictine sustainability can be adapted to changing contexts and timeframes. We also discuss the lessons that may generalize from the Benedictines for practical implications for business. Finally, we conclude with implications and limitations that emerge from this work and suggest research directions to advance knowledge related to sustainability.

4.2 The Rule of St. Benedict

The Benedictines find inspiration for sustainability in the RSB, a historical text written in the sixth century yet actively practiced today in Benedictine organizations worldwide (Mercier & Deslandes, 2017). The RSB is an ancient and timeless document. It is largely as relevant today as when it was written almost 1500 years ago. The ethical foundations for Benedictine communities are found in the RSB as well as a plethora of social, environmental, and economic guidelines. Importantly, the RSB has provided guidance for many centuries yet is currently also applicable to social, environmental, and economic sustainability. This document facilitates a long-term orientation that encourages organizations to notice their long-term impact (Hanson & Keplinger, 2021; Sternad, 2016). Current management practices, particularly in publicly traded organizations, tend to set their goals and initiatives toward shorter-term orientation (Zhang & Gimeno, 2016). Thus, it may be difficult to make a financial case for choosing sustainable practices. Moreover, in for-profit organizations, the focus is on profit which may lead to a failure to focus on social and environmental sustainability or choose sustainability-friendly responses in their business practices unless those sustainability activities lead to economic success (Schaltegger & Hörisch, 2017). Thus, modern organizations may learn from the long-term orientation and longevity of the Benedictines, who faithfully follow the RSB and are self-sufficient and purpose-driven rather than solely profit-driven.

The RSB provides the basics for living a balanced, simple, and values-based life. Monks and nuns commit to follow the edicts of the RSB through their formal vows of stability, conversion, and obedience. In this chapter, we will elucidate how the vow of stability is particularly salient to sustainability. In addition to the vows, the Benedictines share common values described in the RSB that are pertinent to sustainability, including humility, moderation, balance, and living in a community. The value of moderation ensures consumption patterns based on essential needs. The value of humility shifts the focus of attention from the individual to the environment and the community outside oneself. A balanced rhythm of life may foster the well-being of the individuals in a monastic community and may help to sustain the frugal lifestyle for future generations (Hanson & Keplinger, 2021).

Values related to sustainability found in the RSB also align with Christian values, and the RSB often quotes Biblical principles. The first chapter of Genesis speaks about human dominion over nature, which may not always occur in a sustainable way (Dubos, 1972). In contrast, the RSB seems to draw inspiration from the second chapter of Genesis, where humans are placed in the Garden of Eden not so much to master but rather to steward that environment (Dubos, 1972). Benedictines seem to have a relationship with the physical world around them that arises from their values.

The RSB provides behavioral guidelines for Benedictine organizations, written by Benedict of Nursia in the sixth century. The RSB has initially served as the governing document for monasteries he founded in Subiaco and Monte Cassino. The RSB constitutes a basic guide for living the Christian life and continues to be followed by every Benedictine organization in the world today. Benedictine communities refer to the RSB for ethical guidance, but they also augment the teaching of the RSB with modern interpretations of it (Mercier & Deslandes, 2017). Although the RSB is lived by monastics, laypeople and oblates (individuals living in general society, who have individually affiliated themselves with a monastic community of their choice) may choose to follow the values of the RSB without committing to the full vows to live out the vocation of a monk or a nun. Many of the precepts of the RSB are applicable to living outside the monastic community and certainly are applicable for sustainability. In this chapter, we provide three examples of Benedictine organizations and how their practices may generalize to other modern organizations. Other religious orders, which follow the RSB, include the Cistercians and the Trappists; however, Benedictine organizations are unique compared to other monastic organizations in their economic approach as they are self-sustaining economic units. The specifics of the RSB relevant to sustainability and the ethical framework it provides hold great meaning for those seeking to live and work in a sustainable way. Business leaders, business owners, and managers may also choose to follow some of these values or simply learn and implement some the general lessons taught by the Benedictines.

4.3 A Benedictine Perspective on Stability and Sustainability

In recent years, research on sustainability has become prevalent in business and addressed from multiple areas of study such as institutional and ethical perspectives (Fischer et al., 2018; Garriga & Melé, 2004). Sustainability is also studied in disciplines other than business, so it is challenging to find a universal definition of sustainability that encompasses all facets and dimensions. However, generally, scholars agree that primary facets of sustainability are found within the dimensions economic, environmental, and social elements (Elkington, 1994; Fischer et al., 2018). Another important feature tied to sustainability is the importance of time, which is paramount in the related definition of sustainable development described in the next section. Thus, when considering both sustainability and sustainable development, it is important to consider time to reap the long-term benefits of sustainability-driven activities (Bansal & Desjardine, 2014; Fischer et al., 2018).

All the dimensions of sustainability are found in the Benedictine perspective and perhaps most prominently through the vow of stability, one of the three Benedictine vows. The vow of stability obliges all monastics to a long-term commitment of fostering economic stability within their community and working in a sustainable way. The early medieval supply chain differs from the modern one, yet the Benedictine approach remains relevant over time for how they manage goods and resources. While monasteries are viewed as charitable organizations and spiritual centers (e.g., Carroll, 2004), they demonstrate a long-standing history of sustainable business models with expertise in industries such as agriculture, education, forestry, and tourism.

The vow of stability implies sustainability (Rosenberger, 2011) and the idea of stability was established in the RSB thousands of years before our modern conceptualization of sustainability. Monastics, through their vow of stability, are expected to live in such a way that the community can continue to live well in future generations, economically, but more importantly in both humane and spiritual ways (Rosenberger, 2011). Sustainability of the community requires the inner stability of the brothers or sisters. Stability reflects the importance of community and a long-term commitment of those individuals who are part of it. For monastics, this process refers directly to a commitment to the monastery where they will live for the rest of their life. While individuals in business organizations may not vow to stay in their organizations for the whole life, they can still make a clear commitment to be a part of the organization and the business community.

In the Benedictine context, stability is typically associated with a lifetime in one community where the Benedictines are living and working in the same location (Salim, 2021). This vow binds a Benedictine monastic physically and emotionally to the community of profession for the rest of one's life by committing to following the RSB and an Abbot or Abbess as a community leader. Because monasteries are autonomous, the Benedictines commit to living in a specific community despite any hardships and to follow the leadership and practices of that community.

Longoni and Cagliano (2018) found that an organization's time perspective has a critical role in how innovative the organization is in sustainability as well as how much they improve over time. Specifically, their research found that organizations with longer planning timelines, more tolerance for uncertainty, and higher ability levels to learn from past behaviors demonstrate increased sustainable innovativeness. Increasingly, business ethics researchers indicate that it is a moral responsibility for ethical organizations to balance their economic performance alongside social and environmental performance (Florea et al., 2013; Longoni & Cagliano, 2018).

Stability is amenable to reducing individual and organizational carbon footprints. Benedictine life is characterized by little travel by the average monastic as most Benedictines work within the walls of their monastery, as they alternate between work, study, and prayer (Vonk, 2011). For the most part, everything that a monastic needs is present in the monastic community and therefore there is rarely a need to travel.

4.4 Sustainable Development

Medieval monasteries held power as monastics developed previously noble lands to provide food and resources for the poor in sustainable ways (Ehrlich, 2016). The monasteries functioned as intermediaries between the nobles and the poor, serving as pioneers in land and resource management, thus pre-dating concepts such as "environmental justice" and "sustainable development" (Ehrlich, 2016). The term sustainable development was introduced in the 1987 report of the World Commission on Environment and Development, *Our Common Future*, often referred to as the

"Brundtland Report" since that was the surname of the former Norwegian prime minister Gro Harlem Brundtland who was chair of the commission. From this report, the following definition emerged: "Sustainable development is development that meets the needs of the present without compromising the ability of future generations to meet their own needs" (WCED, 1987, 43).

This definition of sustainable development leaves it open to predict the future needs. Scholars view this vague definition as providing both opportunities and challenges in that it does not articulate what needs to be sustained, or how, and does not really explain development; thus, there is an opportunity to extract meaning and build upon this concept (Hedlund de Witt, 2014). However, it is hard to reach agreement about the many ways to interpret sustainable development. The broad nature of the concepts of sustainability and sustainable development as well as the many interpretations presents both opportunities and challenges. On the one hand, there is opportunity in all the potential ways an organization might make a difference, particularly in ecologically responsible ways. On the other hand, the challenge is how to enact sustainable development. This is where Benedictine organizations may provide both historical and modern examples for other organizations, both religious and secular, to emulate. In this chapter, we portray an overview of three examples of sustainability practices in Benedictine organizations.

Sustainable development is often perceived as a part of economic development and associated with less developed countries. However, sustainable development is also important for developed economies. For instance, Hrynkow et al. (2017) show how a Benedictine monastery in a rural province in Canada builds resilience through their economic activity, ecological sustainability, and social responsibility. The monastery, St. Peter's Abbey, provides continuity as the regional economy goes up and down, adapts to meet the needs of the local community, and plays an essential role in rural revitalization.

Freyer et al. (2018) conducted qualitative research on six monasteries consisting of semi-structured interviews and participant observation visits regarding ecological practices in four areas (sustainable/organic agriculture, food consumption, sustainable forestry, and renewable energy). They found that the economics of ecological practices may overpower the decision-making process. In their qualitative results, they reported that the connections between spirituality and ecological practices were expressed by individuals, but not necessarily reflected at the organizational level of the monastery, as only one of the six monasteries established an overall ecological strategy. In theory, the words of the RSB foster sustainable development, but in practice this may be hard to achieve. Despite this, Benedictines aim to sustain operations for future generations, and the RSB gives them guidance. Economic, environmental, and social sustainability and sustainable development are lived practices based upon Benedictine values and beliefs, and the RSB guides and unifies monastic organizations that share that common belief system. The longevity of the Benedictines and their adherence to the RSB is remarkable. From a sustainable development perspective, Benedictines clearly fit the definition as they work toward ensuring continuation of their organizations for future generations and they do so in a sustainable way. In the next three sections, we cover the economic, environmental, and social dimensions from the Benedictine perspective.

4.5 Economic Sustainability

Benedictine organizations follow the following guideline "ora et labora et lege," which means "pray and work and study." Though much of that guideline involves serving in the work of God, it also has economic meaning (Tredget, 2002). Every Benedictine must earn a living for the community to contribute to economic viability. Importantly, Benedictines are committed to future generations which is not the case for all organizations with an economic objective (Ruppenthal, 2019).

The economic approach of Benedictine organizations is entrepreneurial, and they are self-sustaining economic units (Keplinger et al., 2016). Stability, in the monastic context, fosters a long-term perspective (Hanson & Keplinger, 2021). The longevity of monastic communities is not only attributed to religious reasons, as researchers assert that Benedictine monasteries continue to thrive over the centuries due to the establishment of basic governance structures with strong systems of internal and external management and control mechanisms (Feldbauer-Durstmüller et al., 2019). Further, Benedictines are described as pioneers in developing governance mechanisms (Rost et al., 2010).

From an economic perspective, sustainability is fundamental to Benedictine organizations. These communities continue the work of previous generations as well as work toward ensuring the survival of future generations. Thus, a primary economic objective of monastic organizations is to provide financial resources for future generations (Feldbauer-Durstmüller et al., 2019). The RSB describes how Benedictine organizations are to be autonomous operating units (RSB 66:6) which enables these communities to survive despite economically challenging times if they follow the RSB's edict of self-sufficiency and sustainability.

Benedictines often form relationships and partnerships with stakeholders since they need to maintain economic independence and do not receive any financial support from the Catholic Church, and the expectation of the OSB and RSB is economic activity that allows them to self-sustain (Clary, 2007; Ruppenthal, 2019).

4.6 Social Sustainability

Social sustainability is defined as the ability to positively influence the welfare, well-being, and safety of workers as well as sustainable development in the community (Klassen & Vereecke, 2012; Longoni & Cagliano, 2018). Benedictine ethics are focused on both individual and community well-being. The RSB promotes fairness and respect to individual rights as well as fosters inclusion. A Benedictine organization is first and foremost a spiritual institution designed to serve and protect the members, while secondly it is an economic enterprise – at least in the traditional view of profit-driven business organizations (Sedmak, 2014). Overall, there is a clear sense of priorities set forth in the RSB to guide not only spiritual activities but social sustainable behavior for leaders and followers in the organization.

Individual well-being is an important theme in the RSB. Benedictine individuals receive guidance from the RSB on a wide range of daily living activities (e.g., prayer, meals, sleep, work, reading, and intellectual work). These aspects are all covered in the RSB which is quite comprehensive on addressing various aspects of a monastic individual's daily life – both the spiritual and practical sides. There is time allocated in the daily routine for both individual and communal activities in a balanced way which builds resilience (Hanson & Keplinger, 2021). Individual behavior is directed toward building a strong community that operates like a family of brothers or sisters led by either a father (Abbott) or mother (Abbess). The monastics put the well-being of community members first and are directed to love and respect each other in the RSB. Further, guidelines for leaders in the RSB enhance the well-being and personal and professional development of their followers. A social sense of sustainability is found throughout the RSB and is particularly evident in Chapter 64. The monastic leader is directed to take care of monastic brothers or sisters to prevent them from having an overload of work, as well as to encourage personal and professional development of community members.

Interestingly, the environmental benefits of the Benedictine lifestyle may sometimes appear as an indirect effect. Benedictine communities appear to direct their activities choices not so much as directly from environmental values but rather from previously mentioned values that are found in the RSB such as moderation, humility, and putting others first that result in environmentally friendly behavior as described in the next section. Typically, these values lead to behavioral choices that have a relatively low environmental impact while also contributing to the well-being and quality of life of the individual monastics (Vonk, 2011).

4.7 Environmental Sustainability

Environmental sustainability represents a fundamental practice connected with religious values, beliefs, and convictions (Freyer et al., 2018). The ethical foundations for Benedictine environmental practices are also found in the RSB. A core value found in the RSB is humility, and the Latin word for humility, humilitas, originates from the word humus, which means "earth soil." Humility to Benedictines means to have the courage to accept one's own humanity as well as connection and attachment to the earth (Zeitz & Grün, 2013). Humility makes it known that humans are part of creation and should cultivate the earth as stewards rather than owners. Humility also involves accepting one's limitations and taking responsibility for the consequences of their actions (Van Meekeren-Vonk, 2009).

The RSB sets the stage for environmental practices. If designed according to the RSB, a monastery is self-sufficient "where possible, a monastery should always be constructed so that daily needs can be met within the environs of the monastery itself, namely water supply, mill to grind its own flour, and a herb garden, moreover, it should have enough trades and craftsmen to be able to manage its daily running" (RSB 66:6). The rationale is that monastics will "have no need to go running round outside, something which does them no good" (RSB, 66:7).

A case study of the Benedictine Abbey of St Mary in Fulda, Germany, provided evidence that a Benedictine organization uses a business model that is sustainable over time. With a focus on the stewardship and preservation of God's creation, it enables ecological and environmental sustainability (Ruppenthal, 2019). Guided by both the RSB and Christian values, Benedictine organizations clearly demonstrate the environmental pillar of sustainability during their 1500-year history. Environmental sustainability represents a core Benedictine practice connected with religious values, beliefs, and convictions (Freyer et al., 2018). Benedictines are also known for sustainable practices, particularly in agriculture, forestry, food consumption, renewable energy, and tourism (Freyer et al., 2018; Mallarach et al., 2014; Mallarach, 2020). Following Christian values, creation and nature are viewed as sacred. Benedictines are stewards of nature and creation to bestow to future generations (Mallarach et al., 2014; Mallarach, 2020).

While monasteries are often viewed primarily as religious organizations (e.g., Carroll, 2004), they have also developed sustainable models in various industries. As Benedictine communities are strongly guided by Christian ethics and the RSB, their environmental and ecological sustainability (such as organic agriculture and renewable energy production) is also guided by their spirituality and associated ethics. Overall, the Benedictine style of monasticism has positively contributed to ecology over a very long span of time across continents and there is a vast panorama of ecologically positive activity (Mallarach et al., 2014).

Monastics are directed by the RSB to care for all the possessions of the monastery "as if they were sacred vessels of the altar" (RSB 31:10). Possessions of the monastery refers to not only man-made material possessions that facilitate daily living in a monastic community, but these terms also refer to various elements of creation including fields, gardens, forests, and lakes. "Taking care of things" is a virtue held in high regard in Benedictine monasteries (McClure, n.d.). For Benedictines, gardening tools are as important as sacred items used at the altar. This approach from the RSB suggests a connectedness across work, worship, and ordinary activities. The body, material possessions, and the earth are all equivalent and deserving of care and attention. To the Benedictines all things come from God. They have a common-sense approach to their resources, whether that means adopting a new technology or using the same garden shovel for 50 years.

Manual work is fundamentally prescribed in the RSB where it says, "they will be truly monks if they work with their hands" (RSB. 28:8). One of the consequences of this edict and other vows (e.g., stability) and values (e.g., balance) stated earlier is that the Benedictines are tightly integrated with their work. The Benedictines are also very connected to each other, the land, the community. They nurture and care for these interconnections over the long run.

4.8 Examples of Sustainability in Monastic Organizations

The following section provides a few examples of Benedictine organizations practicing sustainability in Europe, Africa, and the USA to illustrate how the economic, social, and environmental pillars are integrated in monasteries on three different continents. These Benedictine communities have common ethical foundations and spirituality, yet they express their long-term commitment to fostering economic stability and diverse social practices and engage in environmental sustainability in unique ways. Together, these examples provide a glimpse of how sustainability is embodied in Benedictine organizations and how sustainable development is fostered.

4.8.1 Admont Abbey

Admont Abbey located in the Ennstal Alps of Austria is a driver of their local economy as the village of Admont has less than 3000 residents (Sternad, 2016). The Abbey was founded in 1074 and has nearly 1000 years of operations, although that was not continuous because there were some disruptions. In 1865, the abbey and the church burned down but were soon rebuilt (Brock, 1907). Later, in the seventeenth and eighteenth centuries, the Benedictines of the Admont Abbey founded a high school and developed the world's largest monastic library (Admont Abbey, 2022). The monastery experienced much harder times in the 1930s. With declining timber prices during the Great Depression, Admont Abbey was economically stressed and liquidated some of their art collection to survive. Just a few years later, in 1939, the monastery was expropriated. Then, World War II occurred, and the monastery was on hiatus until the end of the war.

After World War II, the Admont Abbey diversified. The Admont Benedictines have operated businesses in agriculture, forestry operations, wood panel production, electric power, real estate, and other industries (Sternad et al., 2017). They also run a high school, a nursing home, as well as museums for art and history. The business operations enable the Benedictines to conduct social initiatives and activities that create value in the local community. Today, the abbey's businesses employ approximately 500 individuals (Admont Abbey, 2022). A former Business Director of the Admont Abbey stated, "Our responsibility to our region and sustainable management approaches are important aspects of our work" (Neuner, 2018). This is consistent with the concept of sustainable development as previously described. The long-term philosophy of the Benedictine community clearly comes into play in two respects: the economic sustainability as well as the ecological one.

4.8.2 *Holy Wisdom Monastery*

The Holy Wisdom Monastery in Middleton, Wisconsin, in the USA was originally 130 acres of farmland with few trees, so the Benedictines instituted conservation practices in the early 1970s, restoring acreage to native prairie and wetlands as well bringing the Lost Lake, a 10,000-year-old glacial lake, back to its original size. Later, they realized that the monastery building was extremely energy inefficient, running $100,000 a year in annual heating and cooling costs, and so they decided to make improvements to reduce energy costs as well as improve sustainability (Walgenbach, 2010). "For us, sustainability is not a trend," Sister Joanne Kollasch said on the monastery website, "but a commitment to the earth—a 21st century expression of 1500 years of Benedictine tradition."

Importantly, Holy Wisdom Monastery wanted to show that going green could be done affordably (Walgenbach, 2010). At the time, the newly constructed monastery building achieved the highest LEED Platinum rating ever awarded by the US Green Building Council for a new construction building, as it earned 63 out of 69 points. Walgenbach (2010) stated that monastics were "guided by our mission to weave prayer, hospitality, justice, and care for the Earth into a shared way of life as an ecumenical Benedictine community." This provides an example for modern organizations of Benedictines developed a "business case" for sustainable development based on their mission and values.

4.8.3 *Abbaye Saint-Benoît de Koubri*

After the establishment of the Koubri Abbey in Koubri, Burkina Faso in 1962, the Benedictines introduced various improvements for the local region including electrification and dam construction, improvement of the roads, and the establishment of three new villages close to the Benedictine facilities. Koubri is a twin monastery; the male and female Benedictines are in close spatial proximity and they jointly share land (Langewiesche, 2015). The monks set up a cattle-breeding operation in the 1970s. At the Abbaye Saint-Benoît de Koubri, the Abbess and sisters meet weekly to discuss and decide on any issues concerning the community, such as economic or social activity (Langewiesche, 2015). During the early years of the abbey, the Benedictines have experimented with various businesses until they landed upon yogurt production, which resulted in financial independence. They have a range of business activities from subsistence farming to craft/artisan activities (Langewiesche, 2015).

The abbey with yogurt production and other ventures serves as a reliable employer in the region. International organizations like the World Bank and small NGOs have identified the potential of African monasteries as catalysts for economic development and support monastic projects in the areas of sustainable development, food self-sufficiency, and the protection of biodiversity. While much of the land surrounding the abbey is in production agriculture, the remaining 232 hectares were developed into a biodiversity reserve (Langewiesche, 2015).

There are sometimes tensions between economic and social objectives, as Benedictines are subject to market conditions. Therefore, the sisters in Koubri chose to import subsidized powdered milk from Europe rather than use local fresh milk in the production of their yogurt. This is an example of how profitability goals may conflict with local development strategies (Langewiesche, 2015).

Although all of these examples are of sustainability in religious orders, they could be adopted in secular organizations. Admont Abbey demonstrated sustainability practices that may be widely applied in most any organization. They are community-driven and give back to the local community, providing an example for other organizations. The Holy Wisdom Monastery instituted conservation practices back in the early 1970s, so they have made an impact for 50 years. However, organizations today may choose to follow their lead and institute conservation practices. They also incorporated energy efficiency and reaped cost savings. The latter is particularly attractive for any business seeking to be sustainable while also achieving economic benefits. Finally, Koubri demonstrates an example of ecological sustainability with the biodiversity reserve while adapting their entrepreneurial business models to maintain the self-sufficiency of their organization.

4.9 Discussion

This chapter builds on existing knowledge in numerous ways by providing an overview of very long-standing organizations that effectively manage for sustainability and discussing implications for both research and practice in modern organizations. We show how the Benedictines demonstrated sustainability and sustainable development for centuries and how those concepts are implemented in modern times. Sustainability is a business strategy to meet economic, social, and environmental goals. In discussing the three pillars of economy, environment, and people, we describe the role of sustainability as an umbrella encompassing the interdependency of these concepts. Sustainability requires a complex mutuality of social, economic, and environmental systems. Considering the lived experience with sustainability in Benedictine communities throughout their almost 1500-year history, it is worth looking into their practices. The Benedictine lifestyle serves as a framework for living sustainability following the Christian principles and the guidelines of the RSB.

Connections appear between sustainability, sustainable development, and the ethical and spiritual roots of the Benedictine organizations. We find evidence for sustainability in the RSB and the Biblical foundations on which the RSB is based. However, we cannot reach firm conclusions whether sustainability is tied to the individual level (e.g., monks/nuns, secular employees), monasteries at the organizational level, or both. Freyer et al. (2018) suggests that sustainability efforts originate from the individual rather than organizational level in most cases. However, this research may miss an important aspect found by Vonk (2012), who suggests that

Benedictine communities may not originate their behavior so much from environmental values but rather from the values found in the RSB (e.g., stability, moderation, humility or modesty, the rhythm of life, and reflection). These values, in turn, are expected to influence sustainability.

This chapter certainly has limitations. First, monastic organizations are fundamentally different than secular organizations. Thus, the generalizability of this chapter may be limited. Second, though the RSB suggests a self-sufficient organization, such ideals are challenged by modern times. Some experts criticize that Benedictine monasticism wandered too far from agriculture-based foundations due to changes in farming practices in the 1990s and needs to return to the more sustainable practices (Brown, 2019).

Many Benedictine organizations are struggling with declining populations as their members age. The continuity of these organizations may depend on a new generation of younger individuals committing to Benedictine organizations. There is a strong link between sustainability and spirituality, visible in the RSB and several of the examples of monasteries, which is expected to serve as an important survival factor for Benedictine communities.

Further research is recommended to clarify the degree to which the sustainability themes identified in the current chapter and how these themes resonate with the Benedictines as well as how they view sustainability at both the individual and organizational levels. Further, research could elucidate how Benedictine sustainability influences secular employees in Benedictine organizations as well as impacts other stakeholders. In addition, quantification of the sustainability efforts in the economic, social, and environmental domains would increase and refine our understanding of the role of sustainability in Benedictine organizations and the communities and stakeholders they serve.

4.10 Conclusion

The nearly 1500-year history and current sustainable practices of the Benedictines offer insights to modern businesses. Perhaps the key barrier to transferring Benedictine principles into the workplace is that the Benedictine origins are associated with a religious order and thus may not be acceptable in secular organizations. Though Benedictine organizations are part of a religious order, they also are self-sustaining autonomous value-driven businesses, and many of their underlying values (e.g., moderation) that facilitate sustainability may be applied more widely. The example of Benedictines is especially important as the modern business world strives to understand the meaning and value of sustainability and sustainable development as well as ways of practicing economic, environmental, and social sustainability while considering both the short- and long-term impact of these practices on individuals, organizations, stakeholders, and society. Thus, the 15-century track record of sustainability offered by the Benedictines provides an exemplar to learn how to conceptualize sustainability, live it daily, and continue to practice it over generations. In this chapter, we covered the pillars of sustainability of people, planet (i.e., environmental), and profit (i.e., economic) – the triple-bottom line – from a Benedictine perspective.

References

Admont Abbey. (2022). *History of the Abbey*. Retrieved May 30, 2022 from: https://www.stiftadmont.at/en/the-abbey/monastery/history

Aversano-Dearborn, V., Freyer, B., & Leipold, S. (2013). Sacred sustainability? Benedictine monasteries in Austria and Germany. In International Social Science Council/United Nations Educational, Scientific and Cultural Organization (Ed.), *World social science report 2013: Changing global environments*. OECD Publishing, Paris/UNESCO Publishing.

Bansal, P., & Desjardine, M. R. (2014). Business sustainability: It is about time. *Strategic Organization, 12*(1), 70–78.

Brock, H. M. (1907). Admont. In C. Herbermann (Ed.), *Catholic encyclopedia* (Vol. 1). Robert Appleton Company.

Brown, J. M. (2019). The "greening" of Christian monasticism and the future of monastic landscapes in North America. *Religions, 10*(7), 1–12.

Carroll, A. (2004). *Sustainability and spirituality*. State University of New York Press.

Carroll, A. (2021). Corporate social responsibility: Perspectives on the CSR construct's development and future. *Business & Society, 60*(6), 1258–1278.

Case, S. S., & Chavez, E. (2017). Guiding lights for morally responsible sustainability in organizations: Revisiting sacred texts of Judaism, Christianity, and Islam. *Journal of Organizational Psychology, 17*(2), 35–49.

Clary, B. J. (2007). Broadening the concept of rational economic behavior: A case study of cheese making at the Abbey of Tamié. *Review of Social Economy, 65*(2), 165–186.

Dubos, R. (1972). *A god within*. Charles Scribner's Sons.

Ehrlich, M. (2016, October 6–8). *Environmental justice and sustainable development in the monastery* [Conference presentation]. Southeastern Medieval Association, Knoxville, TN.

Elkington, J. (1994). Towards the sustainable corporation: Win-win-win business strategies for sustainable development. *California Management Review, 36*(2), 90–100.

Elkington, J. (2007). Partnerships from cannibals with forks: The triple-bottom line of 21st-century business. *Environmental Quality Management, 8*(1), 37–51.

Feldbauer-Durstmüller, B., Sandberger, S., & Neulinger, M. (2019). Sustainability for centuries: Monastic governance of Austrian Benedictine abbeys. In *Leadership in the context of religious institutions* (pp. 35–50). Springer.

Fischer, D., Brettel, M., & Mauer, R. (2018). The three dimensions of sustainability: A delicate balancing act for entrepreneurs made more complex by stakeholder expectations. *Journal of Business Ethics, 163*(1), 87–106.

Florea, L., Cheung, Y. H., & Herndon, N. C. (2013). For all good reasons: Role of values in organizational sustainability. *Journal of Business Ethics, 114*(3), 393–408.

Freyer, B., Aversano-Dearborn, V., Winkler, G., Leipold, S., Haidl, H., Brand, K. W., Rosenberger, M., & Wallnig, T. (2018). Is there a relation between ecological practices and spirituality? The case of Benedictine monasteries: Ecological practices in Benedictine monasteries. *Journal of Agricultural & Environmental Ethics, 31*(5), 559–582.

Garriga, E., & Melé, D. (2004). Corporate social responsibility theories: Mapping the territory. *Journal of Business Ethics, 53*(1), 51–71.

Hanson, S. K., & Keplinger, K. (2021). The balance that sustains Benedictines: Family entrepreneurship across generations. *Entrepreneurship & Regional Development, 33*, 442–456.

Hrynkow, C., Ward, B., & Ward, C. (2017). Essential recovery, rural resilience, and energizing communities from St. Peter's Abbey in Muenster, Saskatchewan. *Historical Studies, 83*, 67–88.

Inauen, E., Rost, K., Frey, B. S., Homberg, F., & Osterloh, M. (2010). Monastic governance: Forgotten prospects for public institutions. *American Review of Public Administration, 40*(6), 631–653.

Keplinger, K., Feldbauer-Durstmüller, B., Sandberger, S., & Neulinger, M. (2016). Entrepreneurial activities of Benedictine monasteries-a special form of family business? *International Journal of Entrepreneurial Venturing, 8*(4), 317–333.

Klassen, R., & Vereecke, A. (2012). Social issues in supply chains: Capabilities link responsibility, risk (opportunity), and performance. *International Journal of Production Economics, 140*, 103–115.

Langewiesche, K. (2015). Transnational monasteries: The economic performance of cloistered women. *Social Compass, 62*(2), 132–146.

Laszlo, C. (2008). *Sustainable value: How the world's leading companies are doing well by doing good.* Stanford University Press.

Longoni, A., & Cagliano, R. (2018). Sustainable innovativeness and the triple bottom line: The role of organizational time perspective. *Journal of Business Ethics, 151*(4), 1097–1120.

Mallarach, J. M. (2020). *Monastic communities and nature conservation: Overview of positive trends and best practices in Europe and the Middle East.* Retrieved May 20, 2022, from https://delos-initiative.med-ina.org/wp-content/uploads/2020/08/Monastic-communities-and-nature.pdf

Mallarach, J.-M., Corcó, J., & Papayannis, T. (2014). Christian monastic communities living in harmony with the environment: An overview of positive trends and best practices. *Studia Monastica, 56*(2), 353–391.

McClure, J.M. (n.d.). *About the rule*. Retrieved May 20, 2022, from https://e-benedictine.com/abouttherule/

McDaniel, J. (2002). Spirituality and sustainability. *Conservation Biology, 16*(6), 1461–1464.

Mercier, G., & Deslandes, G. (2017). There are no codes, only interpretations. Practical wisdom and hermeneutics in monastic organizations. *Journal of Business Ethics, 145*(4), 781–794.

Neuner, H. (2018). *Admont Abbey as a driver of the local economy*. https://www.stiftadmont.at/en/enterprises-projects/commercial-enterprises

Order of St. Benedict (OSB). (2022a, May). *A brief history of the Benedictine Order*. Retrieved May 20, 2022, from https://www.osb.org/our-roots/a-brief-history-of-the-benedictine-order

Order of St. Benedict (OSB). (2022b, May). *Monasteries worldwide*. Retrieved May 20, 2022, from http://www.osbatlas.com/

Orr, D. (2002). Four challenges of sustainability. *Conservation Biology, 16*(6), 1457–1460.

Purvis, B., Mao, Y., & Robinson, D. (2018). Three pillars of sustainability: In search of conceptual origins. *Sustainability Science, 14*(3), 681–695.

Rosenberger, M. (2011). Es sollte genügen (RB 39,1; 40,3; 55,4). Elemente eines nachhaltigenafr Lebensstils in der Regel Benedikts LiWiRei. Retrieved May 20, 2022, from https://ku-linz.at/fileadmin/user_upload/Publikationen/div/liwirei_band_2.pdf

Rost, K., Inauen, E., Osterloh, M., & Frey, B. (2010). The corporate governance of Benedictine abbeys – What can stock corporations learn from monasteries? *Journal of Management History, 16*(1), 90–115.

Rule of St. Benedict (RSB). (Original work published in 6th century). L. J. Doyle, Translation into English from Latin (2nd ed., 2001). RSB Online. http://archive.osb.org/rb/text/toc.html

Ruppenthal, T. (2019). The business model of a Benedictine abbey, 1945–1979. *Journal of Management History, 26*, 41.

Salim, E. (2021). Navigating the stay-at-home order with Benedictine stability. *Open Theology, 7*(1), 345–353.

Schaltegger, & Hörisch, J. (2017). In search of the dominant rationale in sustainability management: Legitimacy- or profit-seeking? *Journal of Business Ethics, 145*(2), 259–276.

Sedmak, C. (2014). Sustainability: Ethical perspectives. In C. Weidinger, F. Fischler, & R. Schmidpeter (Eds.), *Sustainable entrepreneurship: Business success through sustainability* (pp. 51–65). Springer.

Sternad, D. (2016). *Long-term orientation in the Benedictine monastery of Admont*. Harvard Business Case Study.

Sternad, D., Kennelly, J. J., & Bradley, F. (2017). *Digging deeper: How purpose-driven enterprises create real value*. Routledge.

Tredget, D. (2002). The rule of Benedict and its relevance to the world of work. *Journal of Managerial Psychology, 17*(3), 219–229.

Van Meekeren-Vonk, M. A. (2009). The quest for religious values and sustainable lifestyles. Contributions form Amish, Hutterite, Franciscan and Benedictine Philosophy of Life. In C. de Pater & I. Dankelman (Eds.), *Religion and sustainable development: Opportunities and challenges for higher education* (pp. 105–127). LIT Verlag.

Vonk, M. A. (2011). *Sustainability and quality of life: A study on the religious worldviews, values and environmental impact of Amish, Hutterite, Franciscan and Benedictine communities* [Doctoral dissertation, Vrije Universiteit Amsterdam]. https://research.vu.nl/ws/portalfiles/portal/42209843/complete+dissertation.pdf

Vonk, M. A. (2012). Sustainability, values and quality of life what we can learn from Christian communities. *Philosophia Reformata, 77*(2), 114–134.

Walgenbach, M. (2010). Doing the right thing: Benedictine women create one of America's greenest buildings. *Aging Today, 31*(4), 5–6.

Witt, H.-d. (2014). Rethinking sustainable development: Considering how different worldviews envision "Development" and "Quality of Life". *Sustainability, 6*(11), 8310–8328.

World Commission on Environment and Development (WCED). (1987). *Our common future*. Oxford University Press.

Zeitz, & Grün. (2013). *The monk and the manager: A discourse on prayer, profits, and principles*. Wiley.

Zhang, & Gimeno, J. (2016). Earnings pressure and long-term corporate governance: Can long-term-oriented investors and managers reduce the quarterly earnings obsession? *Organization Science, 27*(2), 354–372.

Part IV
Confuciansim

Chapter 5
Confucianism and Sustainability

Irene Chu

Abstract Confucianism has much to teach about sustainability, mainly by means of its concept of harmony (*he*), which is probably the most important idea in Confucian culture whilst also arguably the most understudied. It is both a metaphysical and an ethical concept, describing how the world operates and prescribing how human beings should act within it and so has as its goal a "grand harmony," comprised of different components that harmonize as they go through continuous change. This notion of "grand harmony" sets the stage for all other Confucian ideas, being applicable to the triad of heaven, earth, and humanity and so includes the natural world as well as society and individuals. As a moral concept, the principle of harmony is for a person to be a *junzi* (excellent person), who harmonizes but does not seek sameness and should be able to respect different opinions and be able to work with different people in a harmonious way. In this respect, a major function of *li* (correct action) is precisely to achieve this harmony by informing the individual how to act according to their role and to harmonize with others at all levels, those of the individual, society, and the natural world. Subsequently, the concept of achieving harmony has become the goal applicable at all levels of society – not only the individual, familial, organizational, and societal but also at that of the natural environment. This multilevel all-encompassing nature of harmony is the essence of Confucianism's lesson for us all.

Keywords Confucianism · Harmony · Self-cultivation · Naturalistic cosmology · Dynamic vitalism

I. Chu (✉)
School of Management, Newcastle University, Newcastle upon Tyne, UK
e-mail: irene.chu@newcastle.ac.uk

N. Singh et al. (eds.), *Faith Traditions and Sustainability*, Management, Change, Strategy and Positive Leadership, https://doi.org/10.1007/978-3-031-41245-5_5

5.1 Introduction

Confucianism has a great deal to offer to the world in terms of how it can contribute to sustainability (Tucker, 1991), and the main reason for this is its emphasis on harmony (*he*) (Li, 2006, 2008). This concept of harmony does not just include relationships between people; it also extends to include both the natural world and the metaphysical world of heaven – "the great triad of Chinese thought is heaven, earth, and humans" (Tucker, 1991, p. 58). However, these three are not seen as separate and discrete mechanistic parts but should rather be considered as a unified, interconnecting, and interpenetrating whole. There are processes continually operating between these parts of the whole, and so the cosmos is not viewed as being in a state of steady equilibrium but rather in a state of ongoing dynamic change. The Confucian goal of "the grand harmony" is therefore to appreciate this continual change and to work with it to bring all things into balance.

In order to do this, the individual needs to practice self-cultivation, which involves cultivating the virtues of *ren-yi-li* (compassion-appropriateness-correct action) to become a *junzi* or excellent person. However, this is not seen as a purely individual action but is undertaken with an appreciation of the individual's place within a wider world incorporating not only the wider community but also the natural world and heaven. Consequently, the Confucian self can be thought of as a relational self, where the individual exists only in a wider world in which it is relationships with others which define the person. As a result, self-cultivation does not have perfection of the individual as its goal, but rather the emphasis is on helping all others to prosper in the pursuit of the virtues in order to achieve harmony both within society and also with the natural world and heaven.

With such concepts as part of its fundamental way of looking at the world, it is not difficult to see how Confucianism can contribute ideas for a sustainable world. First, because of its natural cosmology of seeing the natural world as part of a trinity with humanity and heaven, there is no disjunction in incorporating it into thinking on how life should be lived. Second, because Confucianism sees the individual automatically as part of society, the extremes of individualism can be tempered. Third, and most importantly, the emphasis on harmony throughout all levels of the society of humanity, earth, and heaven means that it is the balance between things which is considered to be the most important.

While not denying the contribution of other associated traditions such as Daoism and Buddhism, this paper describes how Confucianism, and especially neo-Confucianism and new Confucianism, demonstrates how different moral traditions can make significant contributions to the development of our views on sustainability. To do this, we will first consider the historical development of Confucianism and the environment in which it grew to gain an understanding of its origins. We will then explore the concepts introduced above in more detail before then looking into how these fundamental concepts can be brought to bear on sustainability.

5.2 Historical Roots

Confucian history can be broadly divided into three distinct periods – the time of the Confucian classics or Analects dominated by thinkers such as Confucius or Mencius around 500 B.C.E, the neo-Confucian period in the thirteenth to the fifteenth century, and the new Confucianism dating from the 1920s (Lam, 2003). Each of these has certain characteristics and so we shall briefly consider each in turn.[1]

5.2.1 Classic Confucianism

Confucianism was founded by the sage teacher K'ung Fu-tzu (551–479 B.C.E), whose name was latinized by later Jesuit missionaries as Confucius. He lived at a time of rapid social change when China was undergoing civil war, known as the "Warring States" period with local warlords asserting their strength. Consequently, his aim was to re-establish stability and order through rectification of the individual and the state within a hierarchical social system, and his resulting philosophy emphasized harmony, personal and governmental morality, appropriate social relationships, justice, and sincerity (Yen, 2003). Its foundation was a system of institutional rites and norms that defined and regulated political and social behavior (*li*), previously developed and implemented in the Zhou dynasty (1122–771 BC), and so Confucianism incorporated many common Chinese traditions and beliefs such as strong family loyalty, ancestor veneration, respect for elders, and virtuous behavior. This concept of the family was taken as the basis for government and the principle "do not do to others what you do not want done to yourself" as the golden rule (Lau, 1979). Confucius' singular contribution to this tradition was to provide a philosophical account of the true nature of *li* (correct action) by invoking two quintessential Confucian notions – *ren* (compassion) and *yi* (appropriateness) – as the foundation of its legitimacy. These were coupled with the concept of *hsiao* (filiality) within the central five relations (*wu lun*) – between parent and child, ruler and subject, husband and wife, older and younger siblings, and friends. This provided the basis for the idea that to re-establish order within society, it was necessary to start with harmony within the family. Then, like the concentric circles produced by throwing a pebble into a pond, the effect of virtue would ripple out through society. The fact that this view of harmony was not restricted to human relations but extended to the trinity of mankind, heaven, and earth is shown by the following quote from the *Doctrine of the Mean*:

> Only those who are the most sincere [authentic, true and real] can fully realize their own nature. If they can fully realize their own nature, they can fully realize human nature. If they can fully realize human nature, they can fully realize the nature of things. If they can fully

[1] For more detail, the reader is directed to Lam (2003).

realize the nature of things, they can take part in the transforming and nourishing process of Heaven and Earth. If they can take part in the transforming and nourishing process of Heaven and Earth, they can form a trinity with Heaven and Earth. (Tu, 1989a, p. 77)

Therefore, classical Confucianism can be considered to have a profound environmental ethic, based not only on the triad of heaven, earth, and humans but also on the connecting processes between them.

Confucian thought continued to be developed and refined by his followers after his death (Yen, 2003). Mencius (372–289 B.C.E.) and Hsun Tzu (298–238 B.C.E.) deliberated on whether man was intrinsically good or evil, and Mencius' view of the inherent goodness of human nature went on to give Confucianism its optimistic slant. All these teachings became the basis of the moral tradition of Confucianism, which received official sanction during the Han dynasty (206 BC to 220 AD).

5.2.2 Neo-Confucianism

At about this time, the two religions of Daoism and Buddhism also started to grow in importance, and, while they and Confucianism influenced each other significantly, neo-Confucianism developed as a move to reject their more mystical elements and develop a more rational ethical philosophy. This resulted in a synthesis of earlier teachings by the major neo-Confucian thinker Chu Hsi (1130–1200), recorded in his classic anthology *Reflections on Things at Hand*. Whereas Buddhism proposes a metaphysics of emptiness and withdrawal from the world in mediative practices, neo-Confucianism involves a this-worldly spirituality and a development of one's moral nature to bring it into harmony with the larger changes in the cosmos. This proactiveness in promoting harmony also contrasts with the Daoist view of following the natural way and just letting things be (Li, 2008). These concepts and Chu Hsi's works became the basis of the Chinese civil service examination system in 1315, a system which remained in place for nearly 600 years until 1905.

Another influential neo-Confucian thinker was Wang Yangming (1472–1529), who was living in a time when commercial activity was becoming more important and the merchant class was gaining in power, having previously been ranked at the bottom of the four groups of people after the civil servants, peasants, and industrialists. Wang acknowledged that merchants could become sages if they could harmonize their body and mind (Wang, 1963b) but stressed that human beings must also see heaven and earth as one body by expanding their consciousness to embrace an ever-widening network of relationships.

The great man regards Heaven and Earth and the myriad things as one body. He regards the world as one family and the country as one person. As to those who make a cleavage between objects and distinguish between self and others, they are small men. That the great man can regard Heaven, Earth, and the myriad things as one body is not because he deliberately wants to do so, but because it is natural to the humane nature of his mind that he does so. (Wang, 1963a, p. 659)

In stressing the "humane nature" of the mind as the grounds for the great person to be able to embody the whole universe in their sensitivity, Tu (2001) contends that Wang is making the ontological assertion that this ability to perceive the resonance of heaven, earth, and the myriad of all things is a defining characteristic of being human. He demonstrates this by the following excerpt from the *Inquiry on Great Learning*:

> When we see a child about to fall into the well, we cannot help a feeling of alarm and commiseration. This shows that our humanity (*ren*) forms one body with the child. It may be objected that the child belongs to the same species. Again, when we observe the pitiful cries and frightened appearances of birds and animals about to be slaughtered, we cannot help feeling an 'inability to bear' their suffering. This shows that our humanity forms one body with birds and animals. It may be objected that birds and animals are sentient beings as we are. But when we see plants broken and destroyed, we cannot help a feeling of pity. This shows that our humanity forms one body with plants. It may be said that plants are living things as we are. Yet even when we see tiles and stones shattered and crushed, we cannot help a feeling of regret. This shows that our humanity forms one body with tiles and stones. (Wang, 1963a, pp. 659–660)

Throughout many of the dynastic cycles of China's imperial period, Confucianism was the dominant official ideology and provided significant stability and legitimacy to the ruling regime and to society (Ip, 2011; De Bettignies et al., 2011). As a result, the influence of these concepts of neo-Confucianism on government, education, and society can be considered to be extremely significant (Tucker, 1991).

5.2.3 The Ecological Turn in New Confucian Humanism

This description of the historical roots of Confucianism takes us roughly up to the start of the last century. However, since then there have been further developments – all centered on the comprehensive unity of heaven, earth, and humanity – which Tu (2001) has depicted as an ecological turn. In Taiwan, Qian Mu (1895–1990) stressed the correspondence between this unity and the mutuality of the human heart-mind and the way of heaven (Mu, 1990). In Hong Kong, Tang Junyi (1909–1978) described how the transcendence of heaven is immanent in the communal and critical self-consciousness of humans (Tang, 1977), while in Beijing, Feng Youlan (1895–1990) emphasized the value of harmony not just between humans but also with the natural world (Feng, 1999). Feng was returning to the Confucian ideas he had been developing in the 1940s, prior to the founding of the People's Republic of China. His work is regarded as a representation of the philosophy of harmony as formulated in Zhang Zai's *Western Inscription*, a "core neo-Confucian text in articulating the anthropocosmic vision of the unity of Heaven, Earth, and Humanity" (Tu, 2001, p. 246):

> Heaven is my father and Earth is my mother, and even such a small creature as I finds an intimate place in their midst.
>
> Therefore that which fills the universe I regard as my body and that which directs the universe I consider as my nature.
>
> All people are my brothers and sisters, and all things are my companions. (Zai, 1963, p. 497)

Tu (2001) contends that the work exemplified by these three scholars needs to be considered in the broader context of the world in which they were writing toward the end of the twentieth century. At this time, Taiwan and Hong Kong were undergoing dramatic changes in terms of economic growth and modernization, and even within mainland China, modernization was becoming the most powerful ideology. Consequently, Confucianism may have been seen as being no longer relevant to the contemporary world with its focus on social ethics and family values. However, what these new Confucian thinkers were doing was revitalizing the spiritual aspects of the neo-Confucianism natural organic processes connecting self-cultivation to harmonize with heaven and earth via regulation of the family and of the state.

In some ways, their work echoed that of earlier thinkers such as Xiong Shili (1883–1968) and Liang Shuming (1893–1988) who challenged the move toward Westernization exemplified by the May Fourth Movement in 1919, a key point within the wider New Culture Movement (1915–1921). This was a time when China's humiliation by Western nations and Japan was considered to be caused by the limitations of traditional Chinese culture, including Confucianism, with the proposed solution being based on such Western concepts as the individualism of democracy and the utilitarian ethics of modern science. Xiong Shili argued that it was incorrect for man to impose his will on nature, but man should rather participate in the cosmic and primordial processes of production and reproduction. Human creativity is based on the same vital forces that produce mountains, rivers, and the whole of the planet so that there is consanguinity between humans, heaven, earth, and the multitude of nature (Xiong, 1962). In a similar vein, Liang Shuming emphasized the Confucian ethos of the need for a balance between exploitation and cooperation with nature. He conceded that while it was necessary to take on board Western ideas for the national survival of China, in the long run the need for restraint and moderation in the use of natural resources would become paramount (Liang, 1979).

Tu (2001) concedes that neither of these two scholars or their contemporaries managed to win the argument for Confucian concepts of harmony and sustainability to have precedence over the thrust of the modernist trajectory. However, he notes that "it is now clear that the modernisation process, used simply for the utilitarian ends of development, is insufficient for the full range of human flourishing" (Tu, 2001, p. 251). The resources provided by Confucianism, especially following the ecological turn of new Confucianism, are able to contribute valuably to the drive toward the more sustainable world now being called for.

Confucianism's ongoing relevance is shown by the fact that it continues to have a strong influence on Eastern societies, such as Mainland China, Taiwan, Hong Kong, Singapore, Japan, and Korea. Its wider importance and relevance are shown by the fact that it has been studied from such different fields as philosophy (Ivanhoe, 2002; Sim, 2007; Yu, 2007), cross-cultural research (Hofstede, 1980; Hofstede & Bond, 1988; Franke et al., 1991), business ethics (Chu & Moore, 2020; Ip, 2009b; Romar, 2004a, b), organizational studies and business systems (Whitley, 1999; Hamilton & Biggart, 1988; Redding, 1990), and psychology (Hwang, 2000).

Additionally, Confucianism has often been compared to Aristotelian thought, and it can be considered as a moral tradition (Sim, 2007).

However, in order to further explore how Confucianism as a moral tradition can contribute to our understanding of sustainability, we need to investigate further its essential elements. To do so, we will first look into the main two categories proposed by Tucker (1991), Confucianism's naturalistic cosmology and its ethics of self-cultivation.

5.3 Confucian Elements of Sustainability

5.3.1 Naturalistic Cosmology

The main components of neo-Confucianism's naturalistic cosmology can be characterized as organic holism and dynamic vitalism. In this viewpoint, the universe is not viewed as being composed of discrete mechanistic parts but should rather be considered as a unified, interconnecting, and interpenetrating whole. Everything interacts and affects everything else, with the relationship between the microcosm of oneself and the macrocosm of the universe being essential. This holism is reinforced by the notion of there being no creator God behind the universe, so that "the genuine cosmogony is that of organic process, meaning that all parts of the entire cosmos belong to one organic whole and that they all interact as participants in one spontaneously self-generating process" (Mote, 1971, pp. 17–18). Considerations of how the universe came into being, important for other religions, are less important than the ongoing reality of a self-generating, interconnected universe which has can be considered as a "continuity of being" (Tu, 1989b). This continuity takes the form of a great continual chain of being, an ongoing transformational process connecting the inorganic and all organic lifeforms, including human beings. In neo-Confucian thought, this continuity is reinforced by the fact that "all life is constituted of *ch'i (qi* or energy*)*, the material force or psycho-physical element of the universe" (Tucker, 1991, p. 63), which is then the basis for the deep mutuality of humans and the natural world. This forms the second part of Confucianism's naturalistic cosmology: that of the dynamic vitalism inherent in *qi*.

Dynamic vitalism recognizes that change is the underlying mechanism and basis for interaction between all parts of the universe – mineral, vegetable, and animal including humans. This ceaseless and ongoing process is represented by the term *sheng sheng* (production and reproduction) which reoccurs within neo-Confucian texts. This fundamental process of transformation is to be appreciated and celebrated as an expression of the creative fecundity of life, and humans should seek to harmonize their own actions with these fundamental processes. This viewpoint is vital for a proper understanding of neo-Confucian thought's sense of holism, vitalism, and harmonizing with change which then forms the metaphysical basis for the development of an integrated morality (Tucker, 1991).

5.3.2 The Ethics of Self-Cultivation

Self-cultivation is an important part of Confucianism, emphasizing the notion that a person's moral traits need to be consciously nurtured. The three most important cardinal virtues are *ren* (compassion), *yi* (appropriateness), and *li* (correct action), which act together to constrain self-interest and to promote harmony in a hierarchical social structure (Lin et al., 2013). These provide the *dao* (way) for a person to become good, enabling them to become a *junzi*, translated as an excellent (Yu, 2007) or cultivated person (Yu, 2013), having not only a purposeful life constrained and directed by virtue but also with an essentially social nature, helping others in the pursuit of the virtues in order to achieve social harmony (Lin et al., 2013; Yu, 2007). In this way, Confucianism emphasizes both internal development and external engagement in a continuous process of mutual reinforcement (Ip, 2009b).

However, this striving for harmony should not be thought of as merely involving human relations – it involves harmony with the natural environment as well. It involves a "creative transformation" (Tu, 1985), which can only be understood within the context of the dynamic vitalism of the holistic universe described above. Humans form only one part of the triad together with heaven and earth, and so the resulting humanism is profoundly different from the anthropocentric concept within Western religions – it stresses the unity of man, heaven, and earth rather than the human dominion over nature granted by heaven. Consequently, humans need to live in sustainable harmony with nature, and the notion of exploiting nature is incompatible with Confucian moral self-development (Tu, 1985). Furthermore, this process of self-cultivation should not be seen as separate from the wider cosmological processes of change and transformation. They are inseparable, which is reinforced by the neo-Confucian notion of their being a correspondence between the human virtues and their natural counterparts in cosmic processes. For instance, *ren* – compassion or humaneness – is associated with the principle of origination or fecundity in the wider universe. This is demonstrated by statements made by Chu Hsi such as "humaneness as the principle of love is comparable to a tree and the spring of water" and also "it is the will to grow, like the seeds of peaches and apricots" (Chan, 1969, p. 155). Similarly, the human virtue of *yi* (appropriateness or righteousness) corresponds to the cosmic process of flourishing, *li* (correct action) to advantage, and *zhi* (wisdom) to firmness. In this way, the human virtues are considered to be inseparable parts of the overall one dynamic process of cosmological transformation (Tucker, 1991).

In summary, it can be seen that neo-Confucian thought sees humanity as an integral part of a wider holistic cosmological whole, not static but perpetually transformed by dynamic vitalism. Human moral development by a process of self-cultivation is merely an integral part of these wider universal processes, and so notions of man as a conqueror of nature are simply absurd.

Having now explored these central concepts of naturalistic cosmology and self-cultivation in Confucian thought, we will now proceed to consider how these promote a concept of sustainability which demonstrates the contribution that Confucianism can make to this vitally important area.

5.3.3 Confucian Humanism as an Anthropocosmic Vision

The concept of the unity of heaven, earth, and humans is based on four inseparable levels – those of the self, community, nature, and heaven. The first of these, the self, is not the Western individualistic self but rather a Confucian relational self, whereby individuals see themselves more in terms of their social relations than in the West, reflecting the greater emphasis placed on these social relations by Confucianism. These role-based relationships not only impose obligations and responsibilities on those within them; they also demand that individuals constrain their self-interest to achieve harmony according to the principles of *ren-yi-li*, being compassionate and sincere (*ren*), considering what is the right thing to do (*yi*) and then performing appropriate actions (*li*). This is reflected in the Chinese saying "sacrifice the small self in order to achieve the big self," meaning that personal interest should be constrained in order to achieve the common goal. Confucianism thus emphasizes social relatedness and social roles in the dimension of self-other demarcation and individual identity, and the Confucian individual is interpreted in connection with the community of which he or she is a part (Chan, 2008).

> The self is not an independent entity in Chinese culture; Chinese people have no distinctive awareness of their own existence, uniqueness, direction, goals, or intention. Because there is no clear-cut boundary between oneself and others, the Chinese self can be termed the relational self (Hwang, 2012, p. 192).

This relational self is a very different self from that of the more liberal and individualistic West (Ho, 1995; Yu, 2007; Hwang, 2000) in that whereas the Western self regards the individual as standing alone against the world, the relational self is embedded in a social network. Hwang (2000) sums up the difference between the Confucian and Western concepts of the self well when he refers to the former as being Galilean, whereas the latter is Ptolemaic – i.e., in the West everything revolves around the individual self, whereas in the East, the relational self revolves around the rest of society. Similarly, Fei (1948) likened Western individuals to wooden sticks, which can be bound together into a bundle by social organization, whereas Confucian individuals are more like the ripples caused by throwing a stone into a pond, with each individual at the center of a series of concentric rings. This analogy of concentric rings of ripples in a pond corresponds well to that of the individual situated within the family, then within the face-to-face community, then within the nation, and finally within the whole of humanity. In successively shifting one's empathetic center of concern, a series of challenges can be addressed and overcome. Thus, the move from oneself to the family counteracts selfishness that from the family to community guards against nepotism, from the community to the nation prevents parochialism, and that to the whole of humanity overcomes chauvinistic nationalism (Smith, 1991). However, it is clear that if the focus remains on humanity, this will "undermine our cosmic connectivity and constrain us within an anthropocentric predicament" (Tu, 2001, p. 254). Consequently, a wider focus is required which involves the next level: that of the natural world.

Tu (2001) argues that Confucianism provides the concepts needed for such a sustainable and harmonious relationship between humanity and the natural world while not denying that mankind is "embedded in earth, body, family and community, it never denies that we are in tune with the cosmic order" (Tu, 2001, p. 256). This transcendence is not just an ideal but is manifested in basic Confucian practice. Consequently, the design principle of *fengshui* (wind and water) is based on the alignment of human designs with the environment, enhancing intimacy with nature. Similarly, Chinese medicine takes a more holistic approach than just a narrow treatment of individual symptoms, and breathing exercises known as *qigong* are based on a mutual responsiveness between man and nature. Such notions demonstrate that the Confucian concept of harmony is at the root of this capacity to be able to include the natural world within a transcendent consideration of man's place within the universe.

5.3.4 Confucian Harmony

The concept of harmony (*he*) is probably the most cherished but also undervalued idea in Chinese culture (Li, 2006) and is prominent throughout Confucian teaching up to the present day (Li, 2008). This notion of harmony presupposes the existence of difference and implies the creation of a favorable relationship between disparate elements. Therefore, harmony is not to be understood in the sense of sterile sameness but rather in the sense of bringing together diverse components to make a harmonious whole. For instance, the Confucian classic *Zhu Zhuan* documents the scholar-minister Yan Zi (unknown–500 B.C.E) describing how delicious food cannot be made from similar ingredients but instead requires a blending of different ones:

> Harmony (he) is like making soup. One needs water, fire, vinegar, sauce, salt, and plum to cook fish and meat. One needs to cook them with firewood, mingle (*he*) them together in order to balance the taste. One needs to compensate for deficiencies and reduce excessiveness. The good person (*junzi*) eats [such balanced food] in order to purify his heart/mind (xin). (Li, 2008, p. 424)

Similarly, music cannot be made from just one note but needs a mixture of notes and tunes, some of which may be discordant, but which are resolved to be in harmony with each other:

> Sounds are like flavours. Different elements complete one another: one breath, two styles, three types, four instruments, five sounds, six measures, seven notes, eight winds, and nine songs. Different sounds complement one another: the pure and the impure, the big and the small, the short and the long, the rapid and the measured, the sorrowful and the joyful, the strong and the tender, the slow and the fast, the high and the low, the in and the out, and the inclusive and the non-inclusive. Listening to this kind of music, the heart/mind of the good person (*junzi*) is purified. (Li, 2008, p. 424)

It is not difficult to see how the notion of *he* as the harmonious interplay of sounds or flavors was expanded to other things and become a general theme (Sim, 2007; Li, 2008; Yao, 2000), with Ames (2019) describing it as an "optimal symbiosis."

This concept of Confucian harmony is both a metaphysical and a moral concept. The Chinese word *he* is usually translated as harmony, but a more appropriate translation in some contexts may be "harmonization," reflecting a process rather than an inert state (Li, 2020). At the metaphysical level, the philosophy of *yin yang* (dark and bright) from Daoism became incorporated into Confucianism at the time of the Han dynasty as the process of harmonization, where "female" *yin* elements, such as the moon, water, and darkness, are balanced by "masculine" *yang* elements, such as the sun, fire, and light. The resulting Confucian concept of harmony describes how the entire universe constitutes a "grand harmony," in that it comprises different components that harmonize as they go through continuous change, and this notion of "grand harmony" sets the stage for all other Confucian ideas, being applicable to heaven and earth, the natural world, as well as society and individuals. Thus, harmony is the principle and right path of heaven and "when *yin* and *yang* are balanced, energy (*qi*) flows well, things get generated," and disharmonies are transformed into harmonies (Li, 2008, p. 427).

As a moral concept, the principle of harmony is for a person to be a *junzi* (excellent person), who harmonizes but does not seek sameness and should be able to respect different opinions and be able to work with different people in a harmonious way. In this respect, a major function of *li* (correct action) is precisely to achieve this harmony by informing the individual how to act according to their role and to harmonize with others at all levels, those of the individual, society, and the natural world (Sim, 2007). Subsequently, the concept of achieving harmony has become the ultimate goal and a fundamental value throughout Confucian teaching (Li, 2006). Harmony is the "underlying principle of all relationships, the reason why all virtues can be fully realised" (Yao, 2000, p. 173) and the basic and overlapping goal at different levels – individual, familial, organizational, societal, and at that of the state and the natural environment (Ip, 2009b). Practicing *ren-yi-li* and exercising virtuous acts are the ways to achieve harmonious personal and interpersonal lives. Family and clan patriarchs treating their family members with the virtues produce harmonious families. Rulers practicing *ren-yi-li* and implementing virtue-driven policies create harmonious relationships with their subjects, thus achieving harmonious governance. The undertaking of virtuous acts and policies between states helps to develop a harmonious interstate environment conducive to peace and prosperity for all humanity (Ip, 2009a), and there should also be harmony between human society and the natural world. The ultimate goal is to achieve the "grand harmony" throughout the universe as described above (Li, 2006; Lam, 2003). This is clearly important for the relationship between humans and the natural environment. Due to the fundamental triad of heaven, humans, and the natural world and the continuity of being between them, humans are not entitled to solely exploit natural resources but rather have an obligation to ensure their sustainability.

5.4 The Confucian Harmony Approach to Sustainability

Of course, since the publication of the UN Sustainability Goals, it is accepted that sustainability is not restricted to just the natural environment but also extends to include human, social, and economic sustainability (Brundtland, 1987). How the Confucian notion of harmony can contribute to these wider concepts of sustainability has been a theme investigated by several modern scholars. For instance, Lam (2003) examined how Confucianism developed at different stages throughout Chinese history, from classical Confucians in an agrarian economy through to modern Confucians in the global economy, and summarized that Confucianism's emphasis on harmony promotes equality and sustainability and so can act as a counterweight to some of the extremes of capitalist economic development. This proposition has been further developed by Wang and Juslin (2009). They contend that Western concept of corporate social responsibility (CSR) has largely failed to make a significant impact in the Far East because it cannot be legitimately interpreted within traditional Chinese culture and because the primary reason for CSR is seldom articulated adequately. In response, they propose a reinterpretation of Western CSR using Confucian concepts which is more directly applicable to Chinese culture.

The concept of social responsibility in business in China can be first seen as exemplified by Zi Gong (520–475 B.C.E.), who has been regarded as the original "Confucian Trader" (Lee, 1996). This refers to the fact that he applied Confucian concepts, such as *yi* (appropriateness) and *xin* (sincerity), to his business practices, pursuing profits with integrity and contributing significantly to the prosperity of the community (Lee, 1996). Such Confucian values continued to inform business practices throughout Chinese history, but these declined after the Chinese civil war in 1949 along with the status of Confucianism within society as a whole. However, more recently, it has become evident that China's success in economic growth has been accompanied by a series of problems, including environmental degradation, increasing inequality, and labour rights issues (Wang & Juslin, 2009). Together with external pressures from Western multinational purchasers, these have created pressure for improved standards within business, but which the Western concept of CSR is failing to address adequately.

However, the fundamental tenets of Confucianism can still be applied to business as they have been throughout history up to the start of the Chinese civil war in 1949. In summary, harmonious social relations based on *ren-yi-li* (compassion-appropriateness-correct action) can be applied, suitably modified, to organizations in a similar fashion to how they are applied to humans – profit should result from harmonious business (Confucius, Analects 4:5). It is the fundamental nature of this concept and its wider acceptance within Chinese society which is the reason why it can function as the primary reason behind why organizations should practice CSR (L. Wang & Juslin, 2009). In more detail, Confucian harmony rests on the principles of the family as the core unit of society, reciprocal virtuous behavior, and self-cultivation. Each of these can be adapted to business sustainability as will now be described.

5.4.1 The Concept of the Family

The influence of the family within Confucianism has led to it becoming the basic unit of economic action and consumption, and this, together with the distinctive authority structure of the family, is a vital element in the development of Chinese society (Chu & Moore, 2020; Redding, 1990). In particular, traditional patterns of subordination and obedience within families and acceptance of the dominant authority of the father remain influential and play an important role in the structure of authority relations within businesses. Consequently, businesses often attempt to create a family atmosphere among their employees, emphasizing group membership and a "we" identification, in order to further harmonious relations (Chu & Moore, 2020). This has also been the subject of work by Drucker, who depicts the sustainable organization as a social organization of people performing specialized roles (Drucker, 1983; Romar, 2004b). By adapting familial aspects of *wu lun*, not only are the mutual obligations of both company and employee explicit but also the differentiation of people's roles together with the responsibilities that come with these. While Western thought may view the role ethics of Confucianism as authoritarian and limiting people's individuality, it is not to be forgotten that these role ethics also impose obligations and responsibilities on the more senior side of relationships. These can act as a defense for those on the more junior side of the relationship and so temper some of the excesses of companies mistreating their workers.

5.4.2 Reciprocity

Reciprocity is at the core of the golden rule of "do not do to others what you do not want done to yourself" (Lau, 1979). It is embodied in the term *renqing*, which refers to emotional responses to the various situations of daily life that are guided by sets of social norms for getting along with others (Guo, 2001). *Renqing* has three main meanings. First, it indicates the emotional response that an individual has in various situations in daily life; a person who is versed in *renqing* is well equipped with empathy. Second, *renqing* can be considered as the principle of reciprocity, as a resource that an individual can present to another person as a gift in the course of social exchange with implication of mutual obligation. Finally, *renqing* connotes a set of social norms governing social relations in Chinese society, covering how to interact well with other people and maintain good interpersonal relationships (Hwang, 1987; Wong et al., 2007). It can be seen that the Confucian cardinal virtues of *ren-yi-li* embody *renqing*. Consequently, this concept of reciprocity (*renqing*) can be applied to modern business in the sense of win-win relationships and fair competition. Additionally, it can also be extended to aspects of corporate charity and social conduct. It enables an easier understanding of why enterprises should carry out sustainable practices and CSR in that giving something back to society is acting in line with reciprocity.

5.4.3 Self-Cultivation

It has been described above how self-cultivation to become a *junzi* (a cultivated or excellent person) is a central tenet of Confucianism, as expressed in one of the Analects – "Not cultivating virtue, not learning, not being able to take to justice on hearing it, and not being able to change what is not good: these are my worries" (Confucius, Analects 7:3). In terms of the modern business enterprise, this can be related to the learning organization where there is a continual process of corporate learning to become an excellent company. However, this self-cultivation serves not only to further the interests of the company but also imposes obligations on the company to help others to also prosper – whether these others be employees, external stakeholders, or the natural world. As described above, self-cultivation is part of the dynamic vitalism where all things are interconnected by energy flows within a continuity of being. Consequently, the company should see itself as part of this interconnected cosmos where it is clear that the actions of the company impacting on others need to be explicitly considered. These others again range from employees through to external stakeholders and the natural world, and it is only by considering how the interests of all can be kept in harmony that the company can itself prosper in a sustainable fashion.

The notion that Confucianism can provide a good model for Western companies to act sustainably was also supported by the respected management consultant Peter Drucker (Romar, 2004b). Both Drucker and Confucianism view power and the management of interdependent roles and relationships as central ethical issues in human relations and both emphasize authority, leadership, legitimacy, hierarchy, interdependence, and individual ethical responsibility in their analysis of human affairs. Drucker (1939) explored the failure of capitalism, socialism, and fascism to solve the social challenges of industrialization and considered that capitalism's elevation of economic considerations to dominance had led to the destruction of social bonds. In response, management has three main responsibilities – corporate strategy, the harmonious integration of its members into the hierarchical organization, and handling the relationship between corporate function and social responsibility. Business organizations are primarily hierarchical social organizations made up of specialized people carrying out specialized tasks, but the essence of the corporation is social (Drucker, 1983). Drucker suggests that Confucianism is the most appropriate ethical approach applicable to organizations and their activities, and, while some notions such as *wu lun* may need updating, the Confucian concepts of reciprocity, benevolence, sincerity, and harmony are directly relevant to modern-day business organizations (Romar, 2004b). When all the roles and relationships are performed correctly, those within the firm as well as those between firms, and between firms and society, will be organized correctly and mutual benefit and harmony will result (Romar, 2004b; Li, 2006).

5.4.4 Confucian Entrepreneurs

These principles have been put into practice throughout history – as noted above, Zi Gong (520–475 B.C.E.) is considered to be the first Confucian merchant (Lee, 1996). Such Confucian merchants first became significant in numbers during the *Ming* dynasty (1368–1644), when increasing population pressures made it difficult for all Confucian scholars to follow the previous career path in officialdom (Yu, 1996). This occurred notably in the Huizhou prefecture of Anhui province, which had a strong Confucian tradition which led to the creation of merchant manuals emphasizing the importance of the development of a moral character in line with Confucian moral precepts. Jing Yuan-shan (1841–1903) is cited as a prime example of a Confucian merchant who emphasized the importance of morals over profit-making, and such behavior is still evident in present times (Cheung & King, 2004). For instance, Chou and Cheng (2020) have documented how Confucian principles can form the basis of humanistic management where leaders take the interests of all stakeholders into account and also cultivate employees as humanistic agents. Redding (1990) also described how Confucian principles positively influenced the business practices of overseas Chinese business leaders and case studies of the restaurant chain Din Tai Fung also come to the same conclusion (Lin, 2018; Mayer-Schonberger, 2014; Shu, 2017; Yip, 2019; Yu & Tse, 2012). These exemplify how Confucian principles can form valuable guides for businesses to follow sustainable principles.

5.5 Conclusion

Confucianism has a great deal to offer the world in terms of how its central tenets can be applied to promote sustainability. These range from its ontological roots in the anthropocosmic view of the triad of heaven, earth, and humanity, through its concept of dynamic vitalism leading to a continuity of being connected by *qi* (energy) to its ethic of self-cultivation leading to grand harmony between all things. These result not only in a worldview where nature is not excluded from a primary position in ethics and philosophy but also to a concept of harmony which includes all things – ranging from the natural world and human relations to the relationship of business with external stakeholders and internally within the firm. Businesses can then apply these principles not only to create and maintain a harmonious working environment internally but also with wider stakeholders and the natural environment. Admittedly, some concepts such as *wu lun* may need refining to be meaningful to a worldwide audience, but it is hoped that the depiction given in this chapter convinces the reader that such an attempt is worthwhile.

References

Ames, R. T. (2019). *Confucian harmony (he 和) as an optimizing symbiosis.* Paper presented at the international workshop on "harmony as a virtue" at Nanyang Technological University, Singapore.

Brundtland, G. (1987). *Presentation of the report of the World Commission on Environment and Development to UNEPs 14th governing council*. Retrieved from Nairobi, Kenya

Chan, W. (1969). The concept of man in Chinese thought. In W. Chan (Ed.), *Neo-confucianism, etc.: Essays*. Oriental Society.

Chan, G. (2008). The relevance and value of Confucianism in contemporary business ethics. *Journal of Business Ethics, 77*(3), 347–360.

Cheung, T., & King, A. (2004). Righteousness and profitableness: The moral choices of contemporary Confucian entrepreneurs. *Journal of Business Ethics, 54*(3), 243–257.

Chou, W.-J., & Cheng, B.-S. (2020). *Humanistic paradigm in leadership practice – A case study of a Confucian entrepreneur* (Vol. 27, p. 567). Cross Cultural & Strategic Management.

Chu, I., & Moore, G. (2020). From harmony to conflict: MacIntyrean virtue ethics in a Confucian tradition. *Journal of Business Ethics, 165*, 221–239.

De Bettignies, H.-C., Keung Ip, P., Bai, X., Habisch, A., & Lenssen, G. (2011). Practical wisdom for management from the Chinese classical traditions. *Journal of Management Development, 30*(7/8), 623–628.

Drucker, P. (1939). *The end of economic man: The origins of totalitarianism.* The John Day Company.

Drucker, P. (1983). *Concept of the corporation*. New American Library.

Fei, S. (1948). *Rural China*. Observer.

Feng, Y. (1999). *Zhongguo xiandai zhexueshi (history of modern Chinese philosophy).* Guangdong People's Publishers.

Franke, H. F., Hofstede, G., & Bond, M. H. (1991). Cultural roots of economic performance: A research note. *Strategic Management Journal, 12*, 165–173.

Guo, X. (2001). Dimensions of guanxi in Chinese elite politics. *The China Journal, 46*, 69–90.

Hamilton, G. G., & Biggart, N. W. (1988). Market, culture, and authority: A comparative analysis of management and organization in the Far East. *American Journal of Sociology, 94*, S52–S94.

Ho, D. Y. (1995). Selfhood and identity in Confucianism, Taoism, Buddhism, and Hinduism: Contrasts with the west. *Journal for the Theory of Social Behaviour, 25*(2), 115–139.

Hofstede, G. (1980). *Culture's consequences: International differences in work-related values*. Sage Publications.

Hofstede, G., & Bond, M. H. (1988). The Confucius connection: From cultural roots to economic growth. *Organizational Dynamics, 16*(4), 4–21.

Hwang, K. K. (1987). Face and favor: The Chinese power game. *American Journal of Sociology, 92*(4), 944–974.

Hwang, K. K. (2000). Chinese relationalism: Theoretical construction and methodological considerations. *Journal for the Theory of Social Behaviour, 30*(2), 155–178.

Hwang, K. K. (2012). *Foundations of Chinese psychology: Confucian social relations*. Springer Science & Business Media.

Ip, P. K. (2009a). The challenge of developing a business ethics in China. *Journal of Business Ethics, 88*, 211–224.

Ip, P. K. (2009b). Is Confucianism good for business ethics in China? *Journal of Business Ethics, 88*(3), 463–476.

Ip, P. K. (2011). Practical wisdom of Confucian ethical leadership: A critical inquiry. *Journal of Management Development, 30*(7/8), 685–696.

Ivanhoe, P. J. (2002). *Ethics in the Confucian tradition: The thought of Mengzi and Wang Yangming*. Hackett.

Lam, K. (2003). Confucian business ethics and the economy. *Journal of Business Ethics, 43*(1), 153–162.

Lau, D. C. (1979). Trans. Confucius The Analects (Lun Yu). Penguin Books.
Lee, K. H. (1996). *Moral consideration and strategic management moves: The Chinese case* (Vol. 34, p. 65). Management Decision.
Li, C. (2006). The Confucian ideal of harmony. *Philosophy East and West, 56*(4), 583–603.
Li, C. (2008). The philosophy of harmony in classical Confucianism. *Philosophy Compass, 3*(3), 423–435.
Li, C. (2020). Bring back harmony in philosophical discourse: A Confucian perspective. *Journal of Dharma Studies, 2*(2), 163–173.
Liang, S. (1979). *Dongxi wenhua jiqi zhexue (Eastern and Western cultures and their philosophies)*. Wenxue Publishers.
Lin, G. (2018). *Din Tai Fung*. Commonwealth Publishing Group.
Lin, L. H., Ho, Y. L., & Lin, W. H. E. (2013). Confucian and Taoist work values: An exploratory study of the Chinese transformational leadership behavior. *Journal of Business Ethics, 113*(1), 91–103.
Mayer-Schonberger, V. (2014). In L. Jingyi (Ed.), *Din tai Fung – Perfect temperature*. World Culture.
Mote, F. W. (1971). *Intellectual foundations of China*. Knopf.
Mu, Q. (1990). Zhongguo wenhua dui rennei weilai keyou di kongxian [The possible contribution of Chinese culture to the future of humankind]. *United News*.
Redding, S. G. (1990). *The spirit of Chinese capitalism* (Vol. Vol. 22). Walter de Gruyter.
Romar, E. J. (2004a). Globalization, ethics, and opportunism: A Confucian view of business relationships. *Business Ethics Quarterly, 14*(4), 663–678.
Romar, E. J. (2004b). Managerial harmony: The Confucian ethics of Peter F. Drucker. *Journal of Business Ethics, 51*(2), 199–210.
Shu, Z. (2017). Huawei and Din Tai Fung insist they will not become PLCs. *China Times*.
Sim, M. (2007). *Remastering morals with Aristotle and Confucius*. Cambridge University Press.
Smith, H. (1991). *The world's religions: Our great wisdom traditions*. Harper.
Tang, J. (1977). *Shengming cunzai yu xinling jingjie (Life existence and the spiritual realms)*. Xuesheng Book Co.
Tu, W. (1985). *Confucian thought: Selfhood as creative transformation*. SUNY Press.
Tu, W. (1989a). *Centrality and commonality: An essay on Confucian religiousness a revised and enlarged edition of centrality and commonality: An essay on Chung-yung*. SUNY Press.
Tu, W. (1989b). The continuity of being: Chinese visions of nature. In J. Baird Callicott & R. T. Ames (Eds.), *Nature in Asian traditions of thought: Essays in environmental philosophy* (pp. 67–78). State University of New York Press.
Tu, W. (2001). The ecological turn in new Confucian humanism: Implications for China and the world. *Daedalus, 130*(4), 243–264.
Tucker, M. E. (1991). The relevance of Chinese neo-Confucianism for the reverence of nature. *Environmental History Review, 15*(2), 55–69.
Wang, Y. (1963a). Inquiry on the great learning. In C. Wing-tsit (Ed.), *A source book in Chinese philosophy*. Princeton University Press.
Wang, Y. (1963b). *Instructions for practical living, and other neo-Confucian writings*. Columbia University Press.
Wang, L., & Juslin, H. (2009). The impact of Chinese culture on corporate social responsibility: The harmony approach. *Journal of Business Ethics, 88*(3), 433–451.
Whitley, R. (1999). *Divergent capitalisms: The social structuring and change of business systems*. Oxford University Press.
Wong, Y., Leung, T., Hung, H., & Ngai, E. (2007). A model of guanxi development: Flexibility, commitment and capital exchange. *Total Quality Management & Business Excellence, 18*(8), 875–887.
Xiong, S. (1962). *Xin Weishilun (New theory on consciousness-only)*. Guangwen Publishers.
Yao, X. (2000). *An introduction to Confucianism*. Cambridge University Press.
Yen, T. (2003). *Confucius and Confucianism*. The Commercial Press.

Yip, P. (2019). *Din Tai Fung can teach us about sustainable tourism*. South China Morning Post.
Yu, Y. (1996). Business culture and Chinese traditions: Toward a study of the evolution of merchant culture in Chinese history. In G. Wang & S. L. Wong (Eds.), *Dynamic Hong Kong; business and culture* (pp. 222–272). The Hong Kong University Press.
Yu, J. (2007). *The ethics of Confucius and Aristotle: Mirrors of virtue*. Routledge.
Yu, Y. (2013). *Modern Chinese religious philosophy and spirit of merchantman*. J. Lian, B. Chu, Y. Shi, S. Gong.
Yu, F.-L. T., & Tse, P. W. P. (2012). "Right time, right place, and right people" and Chinese entrepreneurship: A case study of Taiwan's din tai Fung dumpling house. *Frontiers of Business Research in China, 6*(1), 96–119.
Zai, Z. (1963). The Western inscription. In W.-T. Chan (Ed.), *A source book in Chinese philosophy*. Princeton University Press.

Part V
Buddhism & Mindful Approaches

Chapter 6
Buddhist Economics: Philosophical Premises and Environmental Policy Implications

George Gotsis

Abstract Buddhism is a heterogeneous and highly diversified faith tradition embodying value and belief systems that share certain commonalities, among which a sensitivity toward, and a concern for, the environment. Buddhist economics comprises a form of implementing these shared values to a contemporary reality dominated by the pursuit of self-interest without considering common good issues. The aim of this chapter is to elaborate on Buddhist economics as a stream of applied reasoning invested with a strong potential to provide humane and compassionate responses to persistent societal challenges. The present study is intended to expand on this perspective by providing a comprehensive framework of assessing the role of Buddhist economics in implementing policy agendas addressing sustainability and environmental responsibility concerns.

Keywords Buddhist economics · Ecological economics · Buddhist philosophy · Human suffering · Economic ethics · Sustainability · Environmental ethics

6.1 Introduction

Buddhism is a highly diversified faith tradition embodying value and belief systems that share certain commonalities, among which a sensitivity toward, and a concern for, the environment. As a major religious tradition, Buddhism involves a potential for strengthening the entwinements between economy, society, and the environment. Employing a Weberian theoretical framework, Borup (2019) argues that Buddhism has played a significant role in the socioeconomic development and cultural evolution of pre-modern East Asian societies. Despite its rich diversity of doctrinal beliefs and institutional forms, this worldview is intended to entail enduring

---

G. Gotsis (✉)
National and Kapodistrian University of Athens, Athens, Greece
e-mail: ggotsis@phs.uoa.gr

N. Singh et al. (eds.), *Faith Traditions and Sustainability*, Management, Change, Strategy and Positive Leadership, https://doi.org/10.1007/978-3-031-41245-5_6

happiness, cessation of suffering, and abiding balance in all areas of life, as well as to build a harmonious society founded on equanimity, kindness, compassion, and reciprocal joy. Nowadays, Buddhists have positioned themselves with respect to market structures, in the dual form of a critique and an engagement. This intersection of spiritual convictions and market involvement is reflected in coping strategies employed to secure the position of Buddhist practitioners in regional and global economic contexts (Brox & Williams-Oerberg, 2017).

It is widely held that Buddhism is an "eco-friendly" religion, one that nurtures an ecological view of cosmos, by placing emphasis on the primordial unity of human beings and the natural world. Integrated with ecological principles, Buddhist practices assume a positive environmental impact (Darlington, 2017). Buddhist philosophy provides permeating insights on human-environment relations, offering proper responses to persistent issues in environmental ethics. The connection between religious and environmental thought is identified in Buddhist accounts of virtues: character traits such as compassion, humility, and equanimity affect the dispositions of a spiritually enlightened person. For instance, moral dimensions of consumerism can be explored in a virtue ethics context, drawing on Buddhist values such as non-harming, compassion, meditative awareness, and skillful means (Kaza, 2018).

Theoretical models of Buddhist view of the economy have been articulated in the extant economic sociology literature (Aktamov et al., 2015). Max Weber's ideas had a considerable influence on rationalizing the sociological approach to Buddhist worldview. According to Weber (1958), Buddhist economic ethics subordinates the daily life to an overarching body of redemption ideas. Unlike the Calvinist salvation principles that entailed a disenchantment process akin to rationalizing the cosmos, Eastern religious convictions appeared to be unable to rationalize the individual lifestyles. Buddhist soteriological teachings were comprehended only by educated monks, not by the often-illiterate laity. For Weber, teaching mystic and esoteric knowledge did not contribute to the development of empirical science, failing to lead to a rational affirmation of the intrinsic value of secular institutional forms. The Weberian legacy offers a view of Buddhist monasticism as otherworldly, implying that any attempt to conceive of Buddhist involvement in the sphere of economy is highly inauthentic, being a deviation from the spirit of doctrinal scriptures.

Weberian interpretation of Buddhism was reframed by Ernst Friedrich Schumacher, a twentieth-century economist who introduced the term Buddhist economics in academic debate. Buddhist lifestyle was now supposed to originate in the traditional belief of the value of minimizing consumption and production, intended to address real, not superfluous human needs (Schumacher, 1973). Buddhist economics comprises a form of implementing these shared values to a contemporary reality dominated by the pursuit of self-interest, without adequately considering common good issues. Schedneck (2019) argues that the cultural logic of Theravāda Buddhism, in particular the economy of merit and concomitant market engagements, needs to be evaluated in the light of the lived religion method of Meredith McGuire (2008), an approach that prevents us from understanding Buddhist conceptions of the economy as entirely disembodied from Buddhist doctrines.

A common trend within the economics and religion field is to adopt a rational choice theory approach to religious phenomena that focuses on the strategic strategies of humans behaving as rational economic men, as *homines economici*. Scholars within this line of reasoning apply microeconomic theories such as supply-and-demand aspects of religious choice in order to explicate decisions of religious individuals and groups (e.g., Gotsis, 2006, 2007; Gotsis & Drakopoulos, 2011). In this model, religious institutions are mostly conceptualized as businesses that seek to attract new consumers in an emerging marketplace of religious goods within a context of ongoing globalization process. By contrast, Buddhism stresses the connectedness of things, of which economics is only one dimension, and the necessity of access to genuine knowledge of reality. Since Buddhism focuses on realities rather than abstract universals, Buddhist economics is hardly amenable to quantitative modeling (Drechsler, 2019, p. 532).

The aim of this chapter is to elaborate on Buddhist economics as a stream of applied reasoning invested with a strong potential to provide humane and compassionate responses to new societal challenges. The present study is intended to expand on this perspective by providing a comprehensive framework of assessing the role of Buddhist economics in fostering social responsibility and sustainability issues. In so doing, we discuss the primary tenets of Buddhist economics and review studies that operationalize this approach to societal problems in the context of societies influenced by Buddhist traditions. Particular emphasis is placed upon practices that reflect the sustainability dimension of Buddhist values enacted through state policies and organizational strategies that envisage a socially responsible as well as environmentally friendly business environment, especially in the global turmoil of unpredictable and persisting challenges. Finally, tentative suggestions concerning policy implications of this mainstream faith tradition are taken into due consideration.

6.2 Buddhist Philosophy: The Foundations of Economic Ethics

Among the various trends in Buddhist tradition, two specific doctrinal streams of thought provide the philosophical foundations of Buddhist economics: the Theravada and the Mahayana traditions.

Theravada Buddhism (literally, *School of the Elders*) is the most accepted name of Buddhism's oldest extant school. The school's adherents, the Theravādins, have preserved their version of the Gautama Buddha's teaching in the Pāli Canon, the only complete Buddhist canon surviving in a classical Indian language. Theravādins have endeavored to preserve the *dhamma* as recorded in their school's texts. In contrast to Mahāyāna and Vajrayāna versions of Buddhism, Theravāda tends to be more conservative in matters of doctrine and monastic discipline.

Theravada Buddhism is construed in accordance with the four Noble Truths. The first Noble Truth is to be comprehended; the second is to be abandoned; the third is to be realized; and the fourth is to be developed. The realization of the third Noble Truth is conducive to the direct penetration of *Nirvana*, the transcendent freedom that stands as the final goal of Buddha's teachings. The last of the Noble Truths, the Buddha's Eightfold Path of ennobling truths (right view, right intention, right speech, right action, right livelihood, right effort, right mindfulness, and right concentration) provides comprehensive counsels on how to develop those qualities in the human heart that must be cultivated in order to achieve the final goal, the supreme freedom and happiness. In modern terminology, the eightfold path is not infrequently viewed as a strategic approach to a holistic socioecological healing of our human condition, which is in a position to incur a deeper transformation of social structures and institutions (Habito, 2022). For instance, hospitality is perceived as a form of generous giving that helps us to liberate from attachment to desires, thus being a fundamental spiritual virtue (Munasinghe et al., 2017). The ensuing eradication of greed and avarice enables individuals to attain equanimity and wisdom culminating in a state of lovingkindness and compassion. Buddhist virtue ethics is pivotal to shaping attitudes informed by a wider conception of the relations between human well-being and environmental concerns (Cooper & James, 2005; Keown, 2007; Terrone, 2014). This intrinsic motivation for sustainability is imbued by sustainability virtues, interdependent personhood, and an inherent concern for the well-being of others, nature, and future beings (Becker & Hamblin, 2021).

Buddhist hermeneutics arose in a process of interpreting canonical texts, doctrines, and practices. In Tibet, commentators and scholars highlighted the continuity with the Indian Nalanda tradition and adopted its constituent principles, whereas in China, the leading principle was the adaptation of Buddhism to the traditional Chinese mentality and culture (Urbanaeva, 2022). The Mahayana Buddhists follow the path of the Bodhisattvas that seeks not only self-liberation but also the liberation of all sentient beings. What makes Mahāyāna Buddhism distinctive from earlier Buddhism is the emphasis on compassionate responses to others' suffering through deep understanding of selflessness and emptiness. Instead of focusing on self-restraint akin to one's spiritual progress, all virtues should be dedicated to acts of enlightenment. We thus dwell in the four divine abodes: lovingkindness (*mettā*), compassion (*karuṇā*), altruistic joy (*muditā*), and equanimity (*upekkha*).

Mahayana Buddhism offers an ontological foundation to Buddhist generosity, which is a cornerstone upon which Buddhist economics is based. Shen (2018) examines three types of gift, namely, the gift of material goods, the gift of no fear, and the gift of teaching Dharma, and more interestingly, the practice of *huixiang* 迴向 (channeling one's merit to others) as elaborated by Huiyuan in the entry huixiang of his *Dasheng Yizhang* (The Meaning of Mahayana Buddhism). This Chinese Buddhist ethics of generosity can enrich the debate on the desirability of a postmodern ethics of generosity, a concept related to sustainability concerns.

Humanistic Buddhism is a new form of Chinese Buddhism initiated by Master Taixu (1889–1947) and elaborated by subsequent spiritual leaders including Masters Yin Shun (1906–2005), Xing Yun (Hsing Yun, 1927–2003), Sheng Yen (1931–2009), and Zhengyan (Cheng Yan, 1937–). Tai Xu criticized the vehement renunciation of worldly affairs and advocated compassionate concern over the mundane world. However, it was Sheng Yen's systematic framing of a socially engaged Buddhism that inaugurated a peaceful social movement highlighting different ethical and environmental issues, despite its inability to tackle with structural societal challenges (Ng, 2018b).

These principles bear important societal connotations. Tideman (2021, p.190) extrapolates on these teachings by commenting on four basic premises:

Life is experienced as suffering.
The experience of suffering originates in our ignorance regarding the true nature of reality.
The cause of suffering can be removed, insofar as the mind is liberated from ignorance.
There is a certain path to achieve freedom from suffering (the "Eightfold Path").

Framing these four premises as a means of overcoming our ignorance of deeper reality requires a descriptive approach to explain the possible ways through which we are in a position to eliminate ignorance through practice. Tideman (2021, pp. 197–201) explores how insights from Buddhist philosophy can be applied to the creation of sustainable economic systems. For instance, the emphasis on the self as a permanent, autonomous entity stems from either delusion (*moha*) and ignorance (*āvijjā*), the failure to perceive the world as it actually is, or craving (*taṇha*) and inwardly attachment (*upādāna*). He then identifies five clusters of ignorance in the realm of economic reasoning that are in need of a deeper transformation to yield new insights on economic organization informed by the virtue of wisdom. Wisdom (*paññā*) in Buddhism is the ability to understand everything in this world, both internal and external of us, as precisely it is. The dominant economic paradigm is grounded on five types of ignorance regarding the ultimate essence of things:

Ignorance regarding humanity as being in isolation from the cosmos
Ignorance regarding material growth as a universal, yet one-dimensional objective
Ignorance regarding the conception of human beings as self-interested agents exhibiting maximizing economic behavior
Ignorance regarding profit as the sole purpose of business activities
Ignorance regarding the market as being a self-correcting, auto-regulating mechanism that is in a position to sufficiently coordinate plans of economic agents

The prevailing economic paradigm is extremely misapprehending reality because it is entrenched in an inherently individualistic and self-centered worldview, which ignores the ultimate interdependence of all natural entities. For instance, the pursuit of material consumption as an end, which is the primary economic goal, cannot be deemed sustainable. This materialistic ethos results not only in human suffering but also in an ongoing degradation of ecosystems exemplified through an unprecedented ecological crisis.

Tideman (2021, pp. 201–207) seeks to find a viable solution to our current economic predicament through a cognitive process of recognizing the specific role that our minds play in shaping economic outcomes, thus suggesting fruitful ways to shift the mindsets that business leaders adopt to a direction dictated by ancient wisdom traditions. Such a solution embodies a more comprehensive view of reality into our policy frameworks, supported by a proper shift in the modes of thinking of business and political leaders. To stabilize our economic systems, we need to comprehend the particular functions of economic reality not as an autonomous subsystem of social action but as a highly interdependent system of value creation. Such a paradigm shift in economic theorizing has two important connotations:

A reconceptualization of our economic models, considering the mindsets we have to adopt in view of aligning them with sustainable economic outcomes.

A process of restructuring economic institutions and policy frameworks. Such a synergistic perspective is expected to shape a broader economic paradigm that enables human thriving, as well as flourishing of all life forms.

6.3 Buddhist Economics: Ethical and Social Implications

6.3.1 The Basic Premises

The Buddhist approach to economics emerged almost 50 years ago to denote an alternative worldview that challenges the underlying assumptions of Western economics. Mainstream Western economic models are primarily grounded on the core assumptions of instrumental rationality, self-interested behavior; profit-maximization, and instrumental use of natural resources. In the Buddhist worldview, this type of instrumentality is detrimental to human fulfillment; it is perceived as a kind of attachment to worldly engagements that unremittingly induce and proliferate human suffering. Buddhism comprises a philosophical system that emphasizes non-attachment to self-serving, egoistic motives, thus enriching the scope of altruistic and prosocial behaviors.

Buddhist economics represents an alternative to the Western economic mindset. Buddhism is centered on want negation and purification of the human character. This stream of thought places an emphasis on elaborating strategies for addressing global social, economic, and environmental issues at both the institutional and personal levels. In so doing, Buddhist economics focuses on community building centered on normative principles such as social justice, stability of social structure, ecological sensitivity, self-sufficiency, and resiliency.

This line of argument is based upon a normative set of behavioral assumptions: interdependency between economic agents as well as between people and earth; awareness of the need for enlightened self-interest originating in human interconnectedness and altruistic motivation, genuine concern for the overall well-being of employees, customers, shareholders, and local communities; and caring attitudes toward the environment. Precisely because of these assumptions, Buddhist economic models are able to secure prosperity in a sustainable business world enabling minimal suffering of all involved in economic activities (Brown, 2015; Brown & Zsolnai, 2018; Essen, 2009; Zsolnai, 2007).

Buddhist economics is generally defined as the systematic study of how to attain given ends by employing the minimum means. According to Zsolnai (2007), Buddhist economics comprises an alternative strategy applied to an economic setting in view of promoting subjective well-being and societal happiness, by simultaneously respecting the environment. Foundations of Buddhist economics remain the principles of right action and right livelihood: the former consists in delineating self-interested economic behavior through the practice of moral precepts, and the latter refers to the four necessities (food, clothing, shelter, and medicine) integral to human existence. These are prerequisites to attain a state of knowledge gained at the experiential level, namely, wisdom (paññā). Living a virtuous life and practicing meditation are viewed as necessary conditions for gaining *paññā*.

Zsolnai (2009) posits that current business models are designed to promote individualistic pursuits entrenched in narrow self-interest. Buddhist economics on the contrary seeks to demonstrate that the mere satisfaction of insatiable desires appears extremely perilous to the attainment of happiness, by impeding individuals to cultivate their innate potential for self-fulfillment. Not unexpectedly, this approach to economic behavior is inimical to that of Western economics which stresses profit maximization, instrumental use of nature, and utilitarian ethics.

Buddhist economics focuses on middle-way approach, which culminates in the Buddhist connotations of the virtue of moderation, implying an equilibrium state of true well-being. In Buddhist economics, the right amount governs the means of production and consumption based on two kinds of human wants: *tanha*, the pleasure derived from satisfying our desires, and *chanda*, the pursuit of genuine well-being (Kittiprapas, 2015). *Tanha*, the craving to indulge in ostentatious life and ethical promiscuity, is reflective of ignorance that considers production as a basic process of satisfying insatiable desires. On the contrary, *chanda* considers production as the means to achieve true well-being: it is directed toward mutual benefit, conducive to effort and action as founded on intelligent reflection. Therefore, only *chanda* determines the socially desirable level of production and more importantly helps us abstain from occupations entailing a decline in quality of life and imbalance in the natural world. In so doing, Buddhist economics endorses neither asceticism nor luxury (Zsolnai, 2018). The essence of Buddhist economics consists in ensuring that economic activity enhances the quality of life.

This middle path worldview of Buddhist economics has a positive effect on balancing production and consumption levels based on these two kinds of human proclivities. It also ensures mindful economic activities viewed as a means to enhancing well-being (Brown, 2015). In a consumerist economic system, such a reasonable constraint imposed upon human craving would entail multiple benefits for the person, the community, and various ecosystems. Buddhism encourages moderate consumption by influencing one's preferences through meditation and moral reflection (Zsolnai, 2016). Buddhist economics capitalizes on the innate value of interconnectedness, of interdependency between communities and states on one hand, and humans and environment, on the other.

King (2016) summarizes the field of Buddhist economics as consisting of primarily two streams of thought. The first is represented by scholars such as

E.F. Schumacher and the Thai monk Payutto drawing on Theravada Buddhism in a typically Thai context, while the second is linked with the works of Shérab Tendar, a prominent Tibetan Buddhist scholar in the contemporary People's Republic of China who employs a scripturally based Mahāyāna and Tantric Buddhist approach to Economics. Both streams of reasoning share the feature of privileging a renewed economic behavior as an effective technique for individual, social, and environmental emancipation.

6.3.2 *Three Proponents of Buddhist Economics*

6.3.2.1 Ernst Friedrich Schumacher

The German émigré Ernst Friedrich Schumacher (1911–1977), one of the main founders of Buddhist economics intellectual movement, was a rather conventional economist, sharing the belief in the beneficial effects of economic growth and linear progress initiated by new developments in science and technology. However, at about the age of 40, he engaged in prolonged self-examination and spiritual quest, which ultimately helped him become highly critical of Western modernity. Developing a great interest in Buddhist spirituality, he traveled to Burma (today Myanmar) in 1955, where he spent 3 months as a United Nations consultant. During this period, his intellectual encounter with indigenous streams of thought that were suspicious of the penetration of Western development ideals into a traditional society made him skeptical on the uncritical reception of modernity by certain non-Western sociocultural contexts.

In evaluating the country's plan for economic development, Schumacher was able to realize the varieties of indigenous cultural discourses, by trying to reconcile his economic thinking with his new philosophical background. He thus juxtaposed the dehumanizing effects of neoclassical economics with the humanizing implications of right livelihood for Buddhists. Commenting on the need for adjusting development ideals and technological innovation to the particularities of Buddhist societal context, Schumacher stressed the inherent worthiness of Buddhist principles such as nonviolence, conscious limitation of needs, and avoiding inducing harm to others and nature. On the contrary, economies based upon abrupt technological intervention exploiting nonrenewable and thereby exhaustible resources entailed detrimental outcomes to long-run sustainability. In Schumacher' view, such an economic ethics was far from being worthy of emulation by other developing nations (Leonard, 2019). In his vein of reasoning, Buddhist economics delineated a sphere of life in which one displays a set of personal values other than maximizing consumption, thus developing human personality accordingly.

Explicating lifestyles in Buddhist countries require different modes of reasoning with respect to the materialistic justificatory basis of mainstream economics. Schumacher contends that based on Buddhist premises, production and consumption become more rational and efficient, given that they are grounded on simplicity and nonviolence. As a result, Buddhist conception of economic organization seeks to meet genuine human needs by optimizing consumption level, whereas Western economics focuses on maximizing utility derived from consuming a bundle of goods available through optimal production. Buddhist economic ethics is founded on the belief that greed, aggression, and delusion are embedded within market financial structures, thus perpetuating the widespread attachment to the ideals of endless growth and capital accumulation that are deemed responsible for current ecological degradation (Magnuson, 2016). Buddhist economic ethics challenges the basic tenets of mainstream economics and proposes alternative moral principles such as minimizing human suffering, simplifying desires, avoiding violence, and exhibiting genuine care and generosity. Despite this emancipatory potential, Buddhist economic ethics is more pertinent to local subsistence communities aiming at self-sufficiency rather than to cases of large-scale economic integration. King (2016) claims that Schumacher construed Buddhist Economics as an ahistorical corpus that was nonmaterialistic and spiritually imbued, and as such, it comprised a nonmodern spectrum of knowledge, behavior, and values.

To meet this criticism, Magnuson (2022) posits that socially engaged Buddhist economics can significantly benefit from recent developments in the social sciences, from the ideas of philosophical pragmatism and institutional economics. Buddhist holistic worldview is in alignment with the ontology elaborated by Schumacher who envisioned economics as incompatible with neoliberal ideology (Butler, 2013). In this respect, the interconnectedness of humans and the world denotes a view of economic exchanges embedded in the overarching context of the natural environment.

6.3.2.2 Prayudh Aryankura Payutto

The study of business and capitalism in the Buddhist tradition remains a fruitful area of academic scholarship (Williams-Oerberg, 2019). Prayudh Aryankura Payutto (b. 1939), known as Venerable Dhammapitaka, was the most widely read Theravāda Buddhist scholar to supplement Schumacher's ideas with scriptural references (Payutto, 1995). Drawing on Schumacher views, Payutto was strongly opposing the value system underlying Western economics. As a discipline based on the quest for objective knowledge, economics was the intellectual corollary of an ongoing specialization and fragmentation of scientific inquiry accompanying the Industrial Revolution era in Western Europe (Payutto, 2010).

In this view, economics operates in conformity to a narrow disciplinary focus that artificially separates several domains of economic activity from other spheres of human life, among which are affective motivation, social interactions, and environmental responsibility. The solution to this economic predicament would be to delve into the Buddha's teachings on virtuous and non-virtuous mental factors to reconstruct economics based on the proper demarcation between *taṇhā* (craving) and *chanda* (dictated by intelligence and wisdom, anticipating genuine well-being). These two opposing types of desire in Theravāda exegetical analysis motivate human activity in accordance with entirely different goals and causal effects. Even though Buddha never focused on purely economic issues, he rejected extreme asceticism and delivered certain teachings on virtuous economic behavior upon which Buddhist economics can draw. Evidently, economy and prosperity are inherently part of Buddhist teaching, as manifested in certain Buddhist movements that legitimize such tendencies by converting ideals of economic transactions into nonmaterial forms of capital, generating, for example, spiritual capital (Borup, 2018).

6.3.2.3 Indo-Tibetan Buddhist Tradition: Shérab Tendar

Despite these fruitful considerations, Theravāda Buddhist economics remained attached to Schumacher's intellectual legacy. An interesting formulation of the scope and prospects of Buddhist economics from outside Schumacher's circle can be identified in the writings of Shérab Tendar, a prominent Tibetan monastic scholar from Qinghai Province in the People's Republic of China who capitalizes on the vast scriptural inheritance of Indo-Tibetan Buddhist tradition. Shérab Tendar elaborates his version of Buddhist economics based on the Mahāyāna and Tantric sūtras, śastras, vinaya, tantras, and their authoritative Indian and Tibetan commentaries familiar to his Buddhist milieu. Unlike Schumacher, Shérab Tendar's concern is to establish the scriptural authority of Buddhist economics, centered on the Buddha's teaching that wealth should be generated through right livelihood. Economics is not explicitly examined in a single Buddhist scripture, yet an amount of references to the centrality of economic behavior in Buddhist soteriology, in particular the renunciation of amassing wealth for its own sake, can hardly be underestimated. Shérab Tendar shares with Schumacher and Payutto the established belief that Buddhist economics incorporates ethical values intended to confront and neutralize the epistemic authority of normative Western economics, as well as the dominant market-driven mechanisms of development and globalization in Southeast Asia.

6.4 Economic Virtue and Sustainable Behaviors

Buddhist business models are embedded in a broader spiritually inspired framework based on the intrinsic motivation of economic actors to serve the common good: In these models, profits and economic growth are not viewed as ultimate ends in themselves, but only as integral to an overarching perspective imbued by spiritual virtues. Accordingly, cost-benefit analysis cannot be considered the primary criterion of economic decision-making because decision process is integrated in a comprehensive worldview of thinking that favors nonviolence, caring, and generosity (Zsolnai, 2018).

Buddhist virtue of mindfulness could narrow an ethics gap typical of neoclassical economics, the latter often failing to affirm the dimension of human well-being through other than hedonistic lenses. Armstrong (2021) explores how mindfulness practices are employed to harness excesses in consumption: mindfulness substantially helps consumers to ethically constrain their preferences, not uncritically succumbing to advertising. For instance, consumers' compliance to clothing patterns dictated by fashion epitomizes another type of human suffering from a Buddhist standpoint. Fischer et al. (2017) contend that mindfulness fosters sustainable consumption by reducing the propensity for unconscious choices; lowering inattentiveness; refraining from greed and delusion; as well as by increasing prosocial behavior via compassionate practices. Buddhist economics adopts a holistic approach, treating consumers less as prone to acquisition of superfluous goods and more as human beings situated in a certain socioeconomic context.

Through its emphasis on human thriving, Buddhist economics is primarily a human development paradigm (Saengsakorn, 2018). Humans have to meet their basic needs, and these needs must be addressed through the provision of goods and services. Reinert (2018) examines the importance of basic goods in Buddhist economic ethics, suggesting that basic goods provision is integral to Buddhist economics and deserves further attention in ethical reasoning. More specifically, evaluating the ethical quality of economic activity necessitates a process of identifying the type of human desire that underlies each productive engagement. Economic activities supportive of *tanha*, of a craving for immediate gratification, are not only deemed as intrinsically unethical, but they also erode true well-being. This principle applies not only to the private vices but to any economic activity. In sum, wisdom and moderation are required when undertaking decisions with respect to sustainable choices. A healthier economic system has to be evaluated *on the grounds of the moral quality of desire that motivates economic behavior*, as well as by the prospective harmony between all stakeholders, individuals, firms, civil society, and the cosmos.

6.5 Operationalizing Buddhist Philosophy Through Buddhist Economics: Organizational, Societal, and Environmental Policy Implications

6.5.1 Disseminating a Socially Responsible Management Philosophy

As discussed earlier, Buddhist economics captures the economic system as intertwined with other natural and social systems. In this respect, we conceive of economics as a cognitive field that embodies fundamental social, moral, and environmental values in a way that makes economic agents more responsible and accountable for their choices. Drawing on Schumacher's Buddhist economics, Dillard (2009) considers the possibility of developing an integrated perspective on accounting that can be effective in instilling morality to business processes, thus performing a substantial paradigm shift toward a holistic conceptualization of accounting practices and accountability ethics. From the viewpoint of marketing philosophy, Mutakalin (2014) advocates Buddhist economics as an exemplary model of shaping a new consumer ethics centered on refraining from the vicious circle of indulging in satisfying our insatiable desires. Based on Buddhist teachings, such ethical consumption practices comprise activities that avoid incurring harm on other people, animals, or the environment, thus elevating Buddhism to an effective marketing factor that shapes ethical decisions (Ratnayake & Jayawickrama, 2016).

Buddhist concepts are useful to leaders striving to increase organizational performance by implementing practical wisdom from a Buddhist perspective. Van den Muyzenberg (2014) draws on Buddhist wisdom concepts selected together with the scholarly monk Tenzin Gyatso, the 14th Dalai Lama with profound knowledge of Tibetan Buddhism, as well as with the abbot of the Nyanavesakavan temple, P.A. Payutto, one of the most brilliant scholars in Theravada Buddhist economics, as argued earlier. Buddhist concepts from both Theravada and Tibetan traditions can help reduce conflicts, increase subjective well-being by influencing intrinsic motivation, and strengthen self-discipline. However, identifying Buddhist wisdom notions that can be applied to management development remains a difficult endeavor to realize, given the abundance of textual scriptures. This fact requires proper selection processes of which virtues and practices can be assimilated and internalized by contemporary managers, especially in non-Buddhist workplace contexts.

Among the various Buddhist virtues, compassion is a major virtue that can be applied to organizational settings. Atkins and Parker (2012) suggest that psychological flexibility (mindfulness combined with values-directed action) positively influences the perceptual, cognitive, affective, and behavioral aspects of compassion. Yet, in mainstream economic philosophy, compassion appears to be counterproductive, given the preponderance of utilitarian evaluation of economic efficiency, innovation, and performance. In contrast to these prevailing views, Prayukvong and Hoopes (2018) seek to explore the possibility that practicing compassion as the highest virtue can prove beneficial to both corporate performance and sustainability. The authors vividly illustrate an exploratory case study of a Thai firm, "Siam Hands," that exemplifies a Buddhist economics view of social innovation and capacity building, as opposing to conventional utilitarian approaches.

Mindfulness can also yield considerable benefits ranging from enhanced coping mechanisms to address stress to a deeper spiritual enlightenment as experienced by the cultivated mind. Not infrequently however, mindfulness is being degraded to a kind of corporate ideology that serves as an instrument of profit-making rather than as integral to a system of ethics focusing on liberation from suffering. Kovács (2014) explores mainstream and Buddhist economics by identifying potential convergence and divergence between these streams of thought. In so doing, the author contrasts neoclassical to Buddhist economics and provides a robust basis of Buddhist management approaches. More specifically, applying the virtues of mindfulness and compassion to management practice shapes a new administrative philosophy that proffers convenient solutions to the irresolvable issues of today reality.

Indulgence in consumerism is negatively perceived in the Buddhist tradition in general and in Buddhist economics. Drawing on insights from Buddhist philosophy, Bhar (2018) demonstrates how the moral deliberation of Buddhist philosophy regarding consumption can help individuals become more responsible of their consumption patterns. This moral obligation toward individual appropriation of goods stems from the fact that in accordance with Buddhist tradition, indulgence in gratifying selfish desires is considered a source of pain. Based on this premise, consumerism does not entail life satisfaction: only by invoking the notion of mindfulness we can secure a process toward a sustainable and just society.

Buddhism shares with other religious traditions a potential to significantly affect corporate behaviors reflecting social responsibility principles. Examining the impact of Buddhist and Taoist values and beliefs on corporate tax compliance in China, Wang and Lu (2021) found that firms headquartered in locations with stronger salience of religious identities were more likely to display disciplined tax compliance behaviors. Welford (2007) argues that even though many aspects of Buddhist worldview pertain to recent developments in the field of corporate social responsibility, a more radical approach is required if we wish to give further impetus to sustainable development. In his view, Buddhist economics are invested with a strong potential for this decisive conceptual shift. Prayukvong and Foster (2014) seek to demonstrate that social responsibility objectives can be achieved through Buddhist economics. The authors claim that Buddhist-informed management practice can elevate ethical behavior to a prerequisite of corporate social performance, thus making corporate social responsibility the core driver of business activities in a variety of cultural settings.

Employing a comparative perspective, Oppenheim (2017) explores the values embedded in two major religious traditions, zen Buddhism and Judaism, in order to assess their significance for creating a more holistic, inclusive, and responsible management philosophy. In this view, socially engaged Buddhism, exemplified by the Vietnamese monk Thich Nhat Hanh (1926–2022), employed the concepts of mindfulness and interconnectedness to denote the particular ways in which all global issues are intertwined, as well as to highlight our universal responsibility toward our natural environment.

6.5.2 A Focus on Environmental Responsibility

Environmentally friendly attitudes presuppose a set of moral imperatives that underlie an inclusive management approach. Buddhist environmental ontology denotes a deep transformation of the way we think, feel, and act within a relational community of life. Cheng (2016) posits that the central ideas of Mahayana Buddhism can give further impetus to current debate on environmental ethics. In this respect, the Dharma is situated at a specific context of political ecology that delineates the so-called *Ecodharma*, a synthesis of economic and ecological perspectives that substantially broaden the scope of more conventional approaches. This perspective requires a profound transformation of the prospects of economic institutions to enable positive change through explicit environmental sensitivity. Individuals affect their environment through their moral responsibility and moral conduct. This paradigm shift necessitates responsible decision-making procedures entrenched in the principles of normative rationality, reverence, and respect toward stakeholders: individuals, local societies, and the environment.

In Japan for instance, Buddhist environmentalism manifests through various forms of social activism capitalizing on basic Buddhist ideas among which aversion to greediness, interdependence, and views on universality. Dessì (2013) argues that contemporary Japanese Buddhism is increasingly engaging in environmental actions. Despite the emphasis placed on pure religious concerns anchored in a long-standing doctrinal heritage, Japanese Buddhism seeks to reassert its presence in the public sphere, in environmental issues, as well as in cultural dynamics related to the interaction between global uniformity and local particularity.

Several Thai Buddhists seek to clarify and explicate the causes of environmental degradation by suggesting different strategies informed by a set of religious beliefs, practices, and commitments. Seeger (2014) examines concepts of Buddhist environmental ethics as proposed by the most influential proponent of Buddhist economics, Phra Payutto, related to shifting meanings in Thai conceptions of nature. Worthy to mention is the use of metaphors on conceptions of the forest as a spiritual "training ground," as well as of karma as motivational factors for undertaking environmental protection.

Daniels (2014) posits that these Buddhist traditions, grounded in the sub-discipline of ecological economics, are an invaluable resource for both economics and management. More specifically, Buddhist practical philosophy entwined with the emerging paradigm of ecological economics could enrich new management approaches aligned with sustainable and welfare-enhancing economic reforms. This integration of Buddhism with new developments in the field of ecological economics can prove highly beneficial for Buddhist practical ethics in their quest of a sound scientific basis that is in position to challenge the intellectual hegemony of market-oriented economics.

6.5.3 Sustainability and the Need for Economic Transformation

As already highlighted, Buddhist teaching demonstrates that our human condition is imbued with greed, hatred, and delusion, all of which comprise spiritual defilements that result in human suffering. These states are viewed as innate dispositions anchored in institutionalized practices typical of corporate-dominated societies. More specifically, the structural framework underlying mainstream economic reasoning is elevated to an instrument intended to perpetuate economics centered on greed and love of money. This framework is extremely at odds with Buddhist views on economics that embrace a more holistic, ecological vision of economic behavior.

As already argued, Buddhist economics suggests that *pañña*, defined as our innate ability to understand everything in its own nature, should be an invaluable productive resource. Proponents of this stream of thought argue that a state of happiness (*sukha*) is the social corollary of the emergence of *pañña*. Happiness is thus viewed more as a state of inner peace and tranquility rather than as the pursuit of pleasure and immediate gratification of desires which comprise the cornerstone of mainstream economics. Puntasen (2007) claims that Buddhist economics remains an efficient means to achieve proper allocation and use of resources, thus enabling sustainable development processes, especially in times of economic uncertainty. In a similar vein of reasoning, Suriyankietkaew and Kantamara (2019) employ a Buddhist philosophy of sufficiency economy to investigate potential ethical applications. Their findings suggest that, to achieve sustainability, enterprises should embrace spiritual values displaying moderation, reasonableness, resilience, compassion, and genuine care for all stakeholders.

Sustainability is based upon the ideal of self-sufficiency. Chaisumritchoke (2007) draws on the notion of sufficiency economy as a necessary strategy for local pharmaceutical industries in a Theravada Buddhist Thai context. Sufficiency economy as an integral part of Buddhist economics comprises three essential principles: *moderation*, *rationality*, and *self-immunity to changes*. These principles are complemented by two conditions: *wisdom* defined as the ability to conceive of things properly and *morality*, the commitment not to place burden on others. The author introduces three levels of organizational development in accordance with the principle of moderation. At the first level, the local industries are supposed to build their own capabilities by placing an emphasis on internal, rather than external, resources. The second level presupposes a synergistic ideal according to which local companies should engage in mutual support, refraining from competing each other. This collaborative strategy results in network building between local industries.

At the third, societal level, the local pharmaceutical companies should exhibit genuine devotion to serving the needs of Thai people. Insofar as alleviating suffering remains a core strategy of Buddhist economics, the local industries should have to learn from patients' sufferings. Taken for granted that the multinational pharmaceutical companies derive excess profits, local companies would strive to reduce this suffering inflicted upon vulnerable people. Such compassionate policies are pivotal to helping local companies gain support from Thai patients, securing corporate sustainability and reputation in this context.

Buddhist economics is perceived as embodying a coherent set of policy implications, yet there is a variety of particular streams of reasoning on policy interventions regarding sustainability issues. Speece (2019) embeds Buddhist economics into the current debate on sustainability in Thailand by demonstrating the divergence of views on which policies are deemed most effective in attaining sustainability goals. Findings of the study reveal substantial differences between four approaches.

The first perceives of Buddhist economics *as primarily aiming at improving individual moral behavior* in a market economy, thus diminishing the possibility of a major systemic change.

The second appears *as highly critical of modern capitalism*, being supportive of more communal lifestyles pertaining to collectivist, agricultural communities.

The third advocates *an internal reform of existing socioeconomic structures* in favor of a mixed economy incorporating a sort of central planning and state intervention to coordinate market mechanisms.

The last approach to sustainability issues from a Buddhist economics' perspective seeks to avoid extremes *by invoking a middle path emphasizing small business entrepreneurship* and stressing the need for imposing certain institutional constraints on foreign direct investment typical of large corporations.

From another point of view, Song (2020) argues that Thailand developed a unique sustainability framework reflecting a sufficiency economic philosophy, in accordance with the Theravada Buddhist worldview on the interrelationships between society and environment. The author identifies three pillars of management philosophy underlying this framework.

First, moderation dictates actions in conformity to the middle path so as to mitigate self-interest, indulgence in inordinate consumption, and pursuit of individual gratification. Businesses can exist in a symbiotic and synergistic relationship with society and environment by avoiding the two extremes, namely, unlimited growth and economic stagnation.

Second, reasonableness comprises a set of accumulated experience, which enables awareness of the interconnectedness of individual entities. Such an awareness of the interdependency of economic decisions and outcomes proves beneficial to sociocultural and environmental systems.

Third, self-immunity denotes a capability to identify perceived external threats that could be potentially perilous to the proper functioning of the firm. Being mindful of such threats and risks, businesses and local communities can be resilient enough to rapidly address such disturbance factors.

6.6 Discussion

6.6.1 Theoretical Implications

As discussed in the previous sections, ancient Buddhist teaching on environmental sensitivity represents an alternative paradigm, opposed to the reductionist logic of neoclassical economics. The Buddhist concept of happiness, the holistic approach to economic action, the emphasis on right livelihood, as well as on the origin, proper use, and fair distribution of wealth are all imbued with ecological connotations. Wagner (2007) argued that these ancient teachings share common elements with ecological economics. Particular attention should be given to the actual impact of Buddhist economic ethics on the creation of a kind of moral economy underlying the social history of many Asian Buddhist countries.

Buddhism dictates potential solutions to unresolved issues posed by modern societies, helping people to discover an appropriate way of life and healing certain societal problems. With such a focus, Buddhism provides a critique of anthropocentrism, thus contributing to mitigate discrimination, suppression, and competitive mindsets by overcoming selfishness. Fostering experiences of a deeper sympathy between self and others, we are motivated not only to alleviate others' suffering but also to protest social inequalities and overt manipulation and oppression (Takemura, 2019).

Research findings on happiness and economics agenda appear to be consistent with Buddhist economics' analysis, in particular with the fact that increased consumption spending does not necessarily generate life satisfaction and enhance well-being (Drakopoulos, 2008; Frey & Stutzer, 2002). Buddhist worldview can provide skillful means to address this type of discomfort through renewed personal values and philosophy of life (Ash, 2007). Buddhist economics proposes strategies intended to minimize suffering, simplify desires, nurture genuine care for stakeholders, and display generosity in economic affairs. Buddhist economics is in a position to foster ethical attitudes that in turn promote permanent ecological sustainability (Zsolnai, 2011). Buddhist economies of merit, karma, and exchange enable promising ways of understanding the epitome of contemporary capitalism – exchanges in stock markets (Cheung, 2021).

Mainstream economics is devoid of a conceptual equipment for studying human sociality because it treats interpersonal interactions as exogenous variables in economic analysis (Bruni, 2010). On the contrary, Buddhist wisdom, applied to economic decision-making, offers the opportunity for an orderly life based on meaningful and purposeful action. Economic attitudes reflecting ethical promiscuity through conspicuous consumption and ostentatious lifestyle are by no means countenanced in this stream of thought. Acquiring status and power through consumption (Wisman, 2019) reveals the human condition of indulging in inordinate desires, thus remaining far from being considered sustainable mode of life. Even in Western intellectual circles, market economy is conceived as yielding an economic religion based on rampant consumerism as the core value, science as a theoretical base, and renewed worship of material progress as new path of salvation (McCarragher, 2019; Nelson, 2019).

Centered on the ideal of self-sufficiency, Buddhist economics proffers an alternative vision of economic activities that is more likely to lead to human thriving. Instead of assuming humans as rational maximizers, Buddhist economics favors a *homo reciprocans* model according to which men engage in acts of cooperation and sharing. Such a collaborative model is based on expectations about mutually reinforcing behaviors centered on the pursuit of happiness through positive emotions, life engagement, and shared meaning. In this respect, social practices of giving are strengthened through the virtue of generosity that constitutes a core attribute of superior person (*sappurisa*), operating jointly with faith, morality, and wisdom. In the inward disposition to give, Buddhism identifies a psychological motivation inimical to the logic of ceaseless acquisition of goods that proves detrimental to both future generations and the environment.

Buddhist economics differs from conventional economics in many significant aspects. It focuses on the enlightened mind and the ensuing endeavor to understand humans as they are. As a result, Buddhist economics is intrinsically humanistic and genuinely caring for the community and the environment (Ng, 2020, p. 85). The Buddhist perspective substantially broadens the scope of the firm through the organizational metaphor *of a living network* in which value is generated by human relationships inside and outside the organization, based on a shared purpose with society (Tideman, 2021, p. 34). On the contrary, the self-interested and maximizing behavior of the *homo economicus* construct fails to capture the pro-social proclivities and the other-regarding preferences of humans who enter into mutually beneficial, reciprocal relations (Urbina & Ruiz-Villaverde, 2019). Grounded in erroneous perceptions of reality, economic systems should be wisely restructured to alleviate human suffering. These premises can provide new impetus to designing more sustainable institutions as a viable response to current crisis.

Applying Buddhist teachings to economics is expected to alleviate new and emerging societal problems, in particular in times of unprecedented crisis. The application of Buddhist practical wisdom enables processes that can lead to more beneficial management outcomes: it contributes to a broadening of equality and justice pursuits, as well as to innovative solutions on the eradication of persistent poverty by ensuring necessities for vulnerable populations. Humans can experience higher levels of happiness by reasonably restricting excesses in consumption and resource utilization, as well as by displaying compassionate responses toward others and respect for nature. This development process opposes that of mainstream economics which endorses self-centeredness, conspicuous consumption, exorbitant rates of growth and consequently, and unlimited exploitation of resources.

In sum, Buddhist economics allows for an environmentally friendly management praxis based on Buddhist premises discussed throughout this paper. Buddhist economics substantially differs from neoclassical economics in its recognition of ethical issues. In addition to the objective of socially desirable level of production and consumption, Buddhist economics recognizes the intrinsic value of nonfinancial factors such as the process of good thinking, or the constructive use of wisdom among leaders and group members epitomized in right understanding and right thought. Economically efficient activities emerge through networks reminiscent of social capital formation, as well as by employing analytical thinking derived from the experience of continuous learning (Prayukvong, 2005).

6.6.2 Implications for Practice

Buddhism is frequently endorsed in the business ethics literature as a valuable cultural asset shaping dynamic perceptions of self and experiences of meaningful work in organizational contexts (Suriyankietkaew & Kantamara, 2019; Vu, 2022; Vu & Burton, 2022). Buddhist ethical values can substantially enrich the scope of management practices by generating positive experiences at work (Marques, 2012).

Mindfulness, for instance, is an appropriate contemplative practice conducive to improved decision-making processes: Buddhist economics offers considerable avenues for the modern mindfulness movement to situate the development of mental capacities in a context that copes with the emerging challenges to market economies (Ng, 2018a).

Buddhist economics can be implemented in organizational practice. Like most religion-based economics, Buddhist economics is viewed as being not only a mere extrapolation from the basic scriptural writings but also a method that enacts doctrinal truths in specific sociocultural contexts. Examples of that practice include the application of the concept of "Gross National Happiness" in Bhutan, the implementation of a "sufficiency economy" in Thailand, and action-based research on the principle of the Unification of King and People in Yogyakarta, Indonesia (Drechsler, 2019). All these typical cases of applying Buddhist economics' principles in practice share an emphasis on the need for sustainability and, to a more varied extent, on small-scale, traditional farming. However, whereas in Bhutan, for instance, sustainability and environmentalism attained a constitutional rank, in Thailand, sufficiency economics can be identified as probably situated somewhere between neoliberal economics and the addition of certain heterodox elements, being slightly more compatible with liberal agendas.

Lucas (2018) refers to two representative cases, Sarvodaya Shramadana Movement in Sri Lanka and Canon Corporation in Japan, as exemplary enterprises that act in conformity to Buddhist principles. Sarvodaya consists of a network of local communities operating as self-sustaining small economies, the goal of which remains the creation of an economic system that supports sustainable village economies. This type of economic organization is called to meet basic human needs such as clean environment, adequate water supply, minimum clothing requirements, balanced diet, simple housing, basic healthcare and communication facilities, minimum energy requirements, and lifelong education for all. Canon, the prominent Japanese corporation, was founded by a devout Buddhist on the grounds of the Buddhist ethos of serving society, rather than maximizing shareholder profits. This organizational philosophy that culminates in the ideal of promoting the common good significantly affects stakeholders, among which customers, suppliers, competitors, and the natural environment.

Leaders should elaborate on alternative business models in view of incorporating a much broader range of values than those implied in mainstream economics (Stirling, 2014). Explicit inclusion of corporate social responsibility in mission statement and corporate vision is one significant step to this direction. For instance, the virtues of mindfulness and moderation implemented through a substantive corporate code of conduct can help firms secure their financial and socioecological sustainability. In this respect, doing business in the Buddhist way can prove both socially desirable and economically efficient. Moreover, proper responses to consumer demand for goods that are produced fairly, humanely, and sustainably is expected to allow firms to realize the fundamental tenet of doing well by doing good.

6.7 Recommendations for Further Research and Concluding Comments

As argued throughout the paper, orthodox economics fails to recognize the intrinsic value of natural environment. Assessing environmental goods merely on the basis of prices determined by impersonal market forces is an inappropriate model for capturing the complexity of environmental issues. On the contrary, policy decisions on the allocation and use of natural resources must be based on practical reason and accumulated experience.

The nascent field of ecological economics is expected to substantially benefit from Buddhist deeper insights on the ultimate nature of cosmos. Ecological economists share the belief that economics must embody normative values that prioritize the common good over self-interested individual preferences. In so doing, they offer constructive suggestions on how to promote collective initiatives to achieve a socially just sustainability transition, thus opposing the strong demarcation between normative and positive economic theory (Dekker & Remic, 2019; Farley & Kish, 2021; Melgar-Melgar & Hall, 2020; Washington & Maloney, 2020). Buddhist spiritual traditions and ecological economics can provide a realistic value-rationality framework that shapes a unique combination of context-dependent knowledge and common good pursuits in view of directing economic initiatives to the attainment of socially desirable objectives. At the same time, ecological economics can provide more credence to Buddhist insights on sustainable and effective policy interventions (Daniels, 2014).

Radical interpretations of Buddhist economics are also worthy of further consideration. Buddhist economics stresses the need for an internal transformation toward compassionate and charitable acts subject to the assumption that this transformation will manifest in new economic systems. Shields (2018) argues that Mahāyāna Buddhist conceptions of liberation can be significantly expanded through critical reflections on productive labor, established hierarchies, and power structures. He in turn claims that the traditional Confucian virtue of propriety can be combined with the notion of habitus in Pierre Bourdieu and that of home in Schumacher, in view of consolidating the critical and communal dimensions of a more radical approach to Buddhist economics.

Several mainstream religious traditions share common elements on faith-based discourses on sustainability. Acknowledging the intrinsic value of nature requires innovative insights on harmonizing the spiritual and material aspects of economic life. Zsolnai (2017) argues that, despite their different ontological and anthropological premises, contemporary Roman Catholic deep ecology and Buddhist economics highlight that individual gratification of selfish desires is extremely perilous to our very existence in the long run. Rather than based on unlimited capital accumulation, prosperity and happiness should be related to the wholeness of life and the quest for personal meaning. In this respect, identifying a common thread between Buddhist economics and other faith-based economic discourses should broaden our scope and perspectives on responsible and sustainable management practices.

To sum up, in the Buddhist economics worldview, the contemporary societal challenges can be conceived as permanent trends arising out of the increasing interrelatedness between firms, society, and ecosystems. Seemingly irresolvable tensions originating in the opposition between materialistic and post-materialistic needs could be effectively resolved by invoking the principle of innate wisdom identified in Buddhist spirituality. As E.C.H. Ng (2020, p. 190) posits,

> A sustainable transformation to Buddhist Economics is possible by transforming our understanding of the notion of 'I' through a six 'I's strategy: from *individual* to *integral*; from *independence* to *interdependence*; from *ignorance* to *insight*.

Viewed as invaluable spiritual resources, these premises can significantly deepen our understanding of how to design and implement effective sustainability strategies, at micro-individual, meso-organizational, and macro-societal levels.

References

Aktamov, I. G., Badmatsyrenov, T. B., Rodionov, V. A., Nomogoeva, V. V., & Badaraev, D. (2015). Theoretical models of Buddhist economy. *International Journal of Economics and Financial Issues, 5*(3S), 94–99.

Armstrong, C. M. J. (2021). Fashion and the Buddha: What Buddhist economics and mindfulness have to offer sustainable consumption? *Clothing and Textiles Research Journal, 39*(2), 91–105.

Ash, C. (2007). Happiness and economics: A Buddhist perspective. *Society and Economy, 29*(2), 201–222.

Atkins, P. W. B., & Parker, S. K. (2012). Understanding individual compassion in organizations: The role of appraisals and psychological flexibility. *Academy of Management Review, 37*(4), 524–546.

Becker, C. U., & Hamblin, J. (2021). Conceptualizing personhood for sustainability: A Buddhist virtue ethics perspective. *Sustainability (Switzerland), 13*(16), 9166.

Bhar, S. (2018). Consuming with mindfulness: Import of Buddhist philosophy for an ethic toward consumerism. *Pertanika Journal of Social Sciences and Humanities, 26*(3), 1563–1578.

Borup, J. (2018). Prosperous Buddhism, prosperity Buddhism, and religious capital. *Numen, 65*(2–3), 256–288.

Borup, J. (2019). Spiritual capital and religious evolution: Buddhist values and transactions in historical and contemporary perspective. *Journal of Global Buddhism, 20*, 49–68.

Brown, C. (2015). Buddhist economics: An enlightenment approach to the dismal science. *Challenge, 58*(1), 23–28.

Brown, C., & Zsolnai, L. (2018). Buddhist economics: An overview. *Society and Economy, 40*(4), 493–513.

Brox, T., & Williams-Oerberg, E. (2017). Buddhism, business and economics. In M. Jerryson (Ed.), *The Oxford handbook of contemporary Buddhism* (pp. 504–517). Oxford University Press.

Bruni, L. (2010). The happiness of sociality. Economics and Eudaimonia: A necessary encounter. *Rationality and Society, 22*(4), 383–406.

Butler, C. D. (2013). Toward Buddhist economics, building on E.F. Schumacher. In G. R. Ricci (Ed.), *Culture and civilization: Cosmopolitanism and the global polity* (pp. 119–160). Taylor & Francis.

Chaisumritchoke, S. T. (2007). Sufficiency economy, the king's philosophy: An application of Buddhist economics to develop Thai local pharmaceutical industries for sustainable well-being. *Society and Economy, 29*(2), 235–252.

Cheng, C. (2016). Environmental ontology in deep ecology and Mahayana Buddhism. *Environmental Ethics, 38*(2), 145–163.

Cheung, K. (2021). Merit, karma, and exchange: Chinese Buddhist mountain tourism company listings on the stock market. *Journal of the American Academy of Religion, 89*(3), 931–955.

Cooper, D. E., & James, S. P. (2005). *Buddhism, virtue and environment*. Ashgate.

Daniels, P. L. (2014). Practical wisdom for managing sustainable enterprises: Synthesizing Buddhism and ecological economics. *Journal of Management Development, 33*, 797–811.

Darlington, S. M. (2017). Contemporary Buddhism and ecology. In M. Jerryson (Ed.), *The Oxford handbook of contemporary Buddhism* (pp. 487–503). Oxford University Press.

Dekker, E., & Remic, B. (2019). Two types of ecological rationality: Or how to best combine psychology and economics. *Journal of Economic Methodology, 26*(4), 291–306.

Dessì, U. (2013). 'Greening dharma': Contemporary Japanese Buddhism and ecology. *Journal for the Study of Religion, Nature, and Culture, 7*(3), 334–355.

Dillard, J. (2009). Buddhist economics: A path from an amoral accounting toward a moral one. In K. Saravana Muthu & C. R. Lehman (Eds.), *Extending Schumacher's concept of total accounting and accountability into the 21st century* (pp. 25–53). Emerald.

Drakopoulos, S. A. (2008). The paradox of happiness: Towards an alternative explanation. *Journal of Happiness Studies, 9*(2), 303–315.

Drechsler, W. (2019). The reality and diversity of Buddhist economics. *American Journal of Economics and Sociology, 78*(2), 523–560.

Essen, J. (2009). Buddhist economics. In J. Peil & I. van Staveren (Eds.), *Handbook of economics and ethics* (pp. 31–38). Edward Elgar.

Farley, J., & Kish, K. (2021). Ecological economics: The next 30 years. *Ecological Economics, 190*. https://doi.org/10.1016/j.ecolecon.2021.107211

Fischer, D., Stanszus, L., Geiger, S., Grossman, P., & Schrader, U. (2017). Mindfulness and sustainable consumption: A systematic literature review of research approaches and findings. *Journal of Cleaner Production, 162*, 544–558.

Frey, B. S., & Stutzer, A. (2002). What can economists learn from happiness research? *Journal of Economic Literature, 40*(2), 402–435.

Gotsis, G. (2006). Contemporary economic analysis of religion: Problems and prospects (in Greek). *The Greek Review of Social Research, 119/A*(2006), 27–62.

Gotsis, G. (2007). The contribution of rational choice theory to the analysis of religion (in Greek). *Greek Philosophical Review, 24*(2007), 3–25.

Gotsis, G., & Drakopoulos, S. (2011). The economics of the early Christian rhetoric. The case of the second Petrine epistle of the new testament. *History of Economic Ideas, XIX*(1), 17–54.

Habito, R. (2022). Buddhist praxis toward global healing: Cultivating clarity, wisdom, and kinship. *Religions, 13*(4), 315.

Kaza, S. (2018). Buddhist environmental ethics: An emergent and contextual approach. In D. Cozort & J. M. Shields (Eds.), *The Oxford handbook of Buddhist ethics* (pp. 432–452). Oxford University Press.

Keown, D. (2007). Buddhism and ecology: A virtue ethics approach. *Contemporary Buddhism, 8*(2), 97–112.

King, M. (2016). Buddhist economics: Scales of value in global exchange. In M. Juergensmeyer (Ed.), *Oxford handbooks in religion* (online), https://www.oxfordhandbooks.com/view/10.1093/oxfordhb/9780199935420.001.0001/oxfordhb-9780199935420-e-64

Kittiprapas, S. (2015). Buddhist approach and happiness for sustainable development. *The Journal of International Buddhist Studies College, 1*(1), 90–123.

Kovács, G. (2014). The theoretical foundation of Buddhist management practices. *Journal of Management Development, 33*(8/9), 751–762.

Leonard, R. (2019). E.F. Schumacher and the making of "Buddhist economics," 1950–1973. *Journal of the History of Economic Thought, 41*(2), 159–186.

Lucas, M. (2018). The need for and nature of Buddhist economics. In S. Stanley, R. Purser, & N. Singh (Eds.), *Handbook of ethical foundations of mindfulness* (pp. 205–221). Springer.

Magnuson, J. (2016). *From greed to wellbeing: A Buddhist approach to resolving our economic and financial crises*. Policy Press.
Magnuson, J. (2022). *The dharma and socially engaged Buddhist economics*. Palgrave Macmillan.
Marques, J. (2012). Making Buddhism work at work: The transformation of a religion into a seasoned ethical system. *Journal of Management Development, 31*(6), 537–549.
McCarragher, E. (2019). *The enchantments of mammon: How capitalism became the religion of modernity*. Harvard University Press.
McGuire, M. (2008). *Lived religion: Faith and practice in everyday life*. Oxford University Press.
Melgar-Melgar, R. E., & Hall, C. A. (2020). Why ecological economics needs to return to its roots: The biophysical foundation of socio-economic systems. *Ecological Economics, 169*. https://doi.org/10.1016/j.ecolecon.2019.106567
Munasinghe, S., Hemmington, N., Schänzel, H., Poulston, J., & Fernando, T. (2017). Hospitality: Ideologies, characteristics and conditionality in Theravada Buddhism and western philosophy. *Hospitality and Society, 7*(2), 157–180.
Mutakalin, G. (2014). Buddhist economics: A model for managing consumer society. *Journal of Management Development, 33*(8/9), 824–832.
Nelson, R. H. (2019). Economic religion and the worship of progress. *American Journal of Economics and Sociology, 78*(2), 319–362.
Ng, E. C. H. (2018a). Mindfulness and Buddhist economics in the financial market-generating delta or alpha? *Society and Economy, 40*(4), 515–529.
Ng, W. Y. (2018b). Pure land and the environmental movement in humanistic Buddhism. In Y. Wang & S. Wawrytko (Eds.), *Dao companion to Chinese Buddhist philosophy* (pp. 419–440). Springer.
Ng, E. C. H. (2020). *Introduction to Buddhist economics: The relevance of Buddhist values in contemporary economy and society*. Palgrave Macmillan.
Oppenheim, L. H. (2017). Enduring values for contemporary issues: Integrating Buddhist and Jewish morality into contemporary management models. *Philosophy of Management, 16*(1), 55–68.
Payutto, P. (1995). *Buddha dhamma: Natural laws and values for life*. State University of New York Press.
Payutto, P. (2010). *Buddhism and the business world: The Buddhist way to deal with business*. Chandran Publishing House.
Prayukvong, W. (2005). A Buddhist economics approach to the development of community enterprises: A case study from Southern Thailand. *Cambridge Journal of Economics, 29*(6), 1171–1185.
Prayukvong, W., & Foster, M. J. (2014). Buddhist economics meets corporate social responsibility. *International Journal of Economics and Business Research, 8*(2), 175–192.
Prayukvong, W., & Hoopes, J. (2018). A Buddhist economics approach to innovation and capacity building: The case of Siam hands. *Society and Economy, 40*(4), 553–570.
Puntasen, A. (2007). Buddhist economics as a new paradigm towards happiness. *Society and Economy, 29*(2), 181–200.
Ratnayake, N., & Jayawickrama, D. C. (2016). Manifestation of ethical consumption behaviour through five precepts of Buddhism. In N. Ratnayake & D. Chaminda (Eds.), *Leadership and personnel management: Concepts, methodologies, tools, and applications* (pp. 1591–1611). IGI Global.
Reinert, K. A. (2018). The wisdom of need: Basic goods provision in Buddhist economic ethics. *Journal of Buddhist Ethics, 25*, 607–636.
Saengsakorn, R. (2018). Buddhist economics as a human development paradigm. *Journal of Philosophy and Religion, 14*(1), 175–193.
Schedneck, B. (2019). An entangled relationship: A lived religion approach to Theravāda Buddhism and economics. *Journal of Global Buddhism, 20*, 31–48.
Schumacher, E. F. (1973). *Small is beautiful: Economics as if people mattered*. Harper and Raw.

Seeger, M. (2014). Ideas and images of nature in Thai Buddhism: Continuity and change. In B. Schuler (Ed.), *Environmental and climate change in South and Southeast Asia: How are local cultures coping?* (pp. 43–74). Brill.
Shen, V. (2018). The ethics of generosity in Chinese Mahayana Buddhism: Theory and practice. In Y. Wang & S. Wawrytko (Eds.), *Dao companion to Chinese Buddhist philosophy* (pp. 45–67). Springer.
Shields, J. M. (2018). Buddhist economics: Problems and possibilities. In D. Cozort & J. M. Shields (Eds.), *The Oxford handbook of Buddhist ethics* (pp. 407–431). Oxford University Press.
Song, H. (2020). Sufficiency economy philosophy: Buddhism-based sustainability framework in Thailand. *Business Strategy and the Environment, 29*(8), 2995–3005.
Speece, M. W. (2019). Sustainable development and Buddhist economics in Thailand. *International Journal of Social Economics, 46*(5), 704–721.
Stirling, K. (2014). Buddhist wisdom as a path to a new economic enlightenment. *Journal of Management Development, 33*(8/9), 812–823.
Suriyankietkaew, S., & Kantamara, P. (2019). Business ethics and spirituality for corporate sustainability: A Buddhism perspective. *Journal of Management, Spirituality and Religion, 16*(3), 264–289.
Takemura, M. (2019). One step towards Buddhism as philosophy. In G. Kopf (Ed.), *The Dao companion to Japanese Buddhist philosophy* (pp. 71–81). Springer.
Terrone, A. (2014). The earth as a treasure in Tibetan Buddhism: Visionary revelation and its interactions with the environment. *Journal for the Study of Religion, Nature and Culture, 8*(4), 460–482.
Tideman, S. (2021). Buddhist economics. In *Could Buddhism provide the intellectual foundations for designing a sustainable economic system?* Waxmann Verlag.
Urbanaeva, I. S. (2022). The role of Chinese and Tibetan hermeneutic strategies in the "constructing" of the divergent worlds of Mahayana Buddhism. *Religiovedenie, 2022*(1), 25–34.
Urbina, D. A., & Ruiz-Villaverde, A. (2019). A critical review of homo economicus from five approaches. *American Journal of Economics and Sociology, 78*(1), 63–93.
van den Muyzenberg, L. (2014). The contribution of Buddhist wisdom to management development. *Journal of Management Development, 33*(8.9), 741–750.
Vu, M. C. (2022). Sensemaking and spirituality: The process of re-centring self-decentralisation at work. *International Journal of Human Resource Management*. https://doi.org/10.1080/09585192.2021.1991977
Vu, M. C., & Burton, N. (2022). Bring your non-self to work? The interaction between self-decentralization and moral reasoning. *Journal of Business Ethics*. https://doi.org/10.1007/s10551-021-04975-1
Wagner, H. (2007). Buddhist economics: Ancient teachings revisited. *International Journal of Green Economics, 1*(3–4), 326–340.
Wang, J., & Lu, J. (2021). Religion and corporate tax compliance: Evidence from Chinese Taoism and Buddhism. *Eurasian Business Review, 11*(2), 327–347.
Washington, H., & Maloney, M. (2020). The need for ecological ethics in a new ecological economics. *Ecological Economics*, 169. https://doi.org/10.1016/j.ecolecon.2019.106478
Weber, M. (1958). *The religion of India: The sociology of Hinduism and Buddhism*. Translated by Hans Heinrich Gerth and Don Martindale. Free Press.
Welford, R. (2007). Examining, discussing and suggesting the possible contribution and role of Buddhist economics for corporate social responsibility. *International Journal of Green Economics, 1*(3–4), 341–350.
Williams-Oerberg, E. (2019). Introduction: Buddhism and economics. *Journal of Global Buddhism, 20*, 19–29.
Wisman, J. D. (2019). Adam Smith and Thorstein Veblen on the pursuit of status through consumption versus work. *Cambridge Journal of Economics, 43*(1), 17–36.
Zsolnai, L. (2007). Western economics versus Buddhist economics. *Society and Economy, 29*(2), 145–153.

Zsolnai, L. (2009). Buddhist economics for business. In L. Zsolnai, Z. Boda, & L. Fekete (Eds.), *Ethical prospects: Economy, society and environment* (pp. 89–99). Springer.

Zsolnai, L. (2011). Buddhist economics. In L. Bouckaert & L. Zsolnai (Eds.), *Handbook of spirituality and business* (pp. 88–94). Macmillan.

Zsolnai, L. (2016). Buddhism and economic development. In T. Lewis & G. de Angelis (Eds.), *Teaching Buddhism: New insights on understanding and presenting the traditions* (pp. 344–360). Oxford University Press.

Zsolnai, L. (2017). Frugality and the intrinsic value of nature. In O. Jakobsen & L. Zsolnai (Eds.), *Integral ecology and sustainable business* (pp. 67–76). Emerald Publishing.

Zsolnai, L. (2018). Economics beyond the self. In V. M. B. Giorgino & Z. Walsh (Eds.), *Co-designing economics in transition: Radical approaches in dialogue with contemplative social sciences* (pp. 89–97). Palgrave Macmillan.

Chapter 7
Cultivation of Loving-Kindness and Compassion: A Societal Solution to Uphold Strong Sustainability Principles in Ecological Policies

Viet-Ngu Hoang, Tsewang Topden, and Florence Ingram

Abstract Strong sustainability argues that substitutions of human and reproducible capitals for natural capital are very limited; hence, upholding strong sustainability principles in ecological and climate policies is necessary to tackle existing global ecological crisis. Implementing these principles requires the accumulation of altruistic capital which includes altruistic preferences and behaviors in individuals and organizations. In this article, we present a narrative arguing that cultivating loving-kindness and compassion (LK&C) in individuals is key to the accumulation of altruistic capital. More importantly, we provide several logical arguments to support two hypotheses: first, it is possible for individuals to develop LK&C; and second, LK&C in individuals can be cultivated limitlessly. Our analysis is based on the classical cognitive framework of Buddhism through the lenses of Middle Way Consequence philosophy.

Keywords Ecological sustainability · Altruism capital · Cognitive framework · Buddhism philosophy

V.-N. Hoang (✉)
QUT School of Business, Brisbane, QLD, Australia
e-mail: vincent.hoang@qut.edu.au

T. Topden
Sera Jey Monastery University, Bylakuppe, India

F. Ingram
TKSL School of Buddhist Science, Brisbane, QLD, Australia

N. Singh et al. (eds.), *Faith Traditions and Sustainability*, Management, Change, Strategy and Positive Leadership, https://doi.org/10.1007/978-3-031-41245-5_7

7.1 Introduction

A traditional economic approach to natural and ecological environment is to consider it as a form of capital in conjunction with human capital and manmade reproducible capital (Stern, 1997; Victor, 2020). Within this approach, a distinction is made between weak and strong sustainability. Weak sustainability suggests that human and reproducible capital can substitute natural capital; hence, natural capital is not essential in maintaining the well-being of human over generations. Strong sustainability argues that substitutions of human and reproducible capitals for natural capital are very limited, and damages or losses caused by the critical natural environment are irreversible (Ayres et al., 2001).

Historically, economies have applied weak sustainability principles in making decisions about the use of natural environment for the production and consumption of goods and services, which have caused sharp decrease in the quantity and quality of available natural resources. As a consequence, climate change has been accelerated to the level that threatens our human ability to reverse those changes (Nunez et al., 2019; Eyring et al., 2021). In addition, empirical research has strongly rejected strong substitutability between natural capital and reproducible capital (Cohen et al., 2019), leaving human society with no other way than to uphold strong sustainability principles. At least critical natural capital, especially ecological services and resources, needs to be preserved to ensure the well-being of future generations (Daily et al., 2000; Pelenc & Ballet, 2015).

Implementing strong sustainability principles in important policy frameworks including climate change policies at state, national, and international levels is very difficult (Ekins et al., 2003; DesRoches, 2019). In most democratic societies, these policies need to pass through public debates which require economic valuation of these policies. Cost and benefit analysis (CBA) – one of the commonly used monetary valuation frameworks – evaluates costs (loss) and benefits (gains) occurring to economic agents over time and across space and discounts all monetary value flows to the present to yield net present value (Pearce et al., 2006). A positive net present value indicates that the policy under analysis is worthwhile to pursue, meaning that the resources needed to implement the policy would increase the total welfare. Historically, CBA of many climate change policies yielded negative net present value, which implied that the proposed policy was not considered worthwhile to pursue (Lind, 1995). Scholars propose important modifications to existing CBA and how natural resources are valued (Costanza, 2020). These include imposing higher monetary values for environmental resources, larger weights in valuating impacts of the future generations and disadvantaged population groups and using extremely low discount rates for future generations in the flexible discounting schemes (Tol, 2001; Dennig, 2017; Kelleher, 2017).

These proposed modifications are based on altruistic preferences; hence, institutionalizing these modifications requires more altruistic preferences and behaviors of key economic agents. Loving-kindness and compassion (LK&C) are shown to be mental motivators for prosocial preferences and behaviors (Walsh-Frank, 1996;

Simpson & Willer, 2015; Gilbert, 2019). Certainly, enhancing altruistic capital is only possible through ongoing cultivation of LK&C in individuals. This raises two research questions: (1) can individuals generate LK&C and (2) can individuals cultivate LK&C limitlessly?

It appears that there is little discussion on these two questions despite growing literature on LK&C from philosophical and ethical perspectives (Walsh-Frank, 1996; Conway, 2001; Yao, 2006; Martin et al., 2014; White, 2017; Augustine & Wayne, 2019; Makransky, 2021). On the other hand, there are remarkable and significant contributions on the subject by His Holiness, the 14th Dalai Lama (Lama, 2006; Lama & Chodron, 2020, 2021), and other Buddhist monastic scholars (Rabten, 1981; Rinbochay & Napper, 1981; Tsering, 2006; Dunne et al., 2020; Tsongkhapa & Jinpa, 2021). These provide the basis for specific discussions on LK&C. Particularly, these discussions are based on the framework of mind and mental factors – a classic framework for understanding cognition in Buddhist texts.

This chapter uses this cognitive framework to gain a deeper understanding LK&C. Importantly, we provide logical arguments to address the two research hypotheses mentioned earlier. Having sound logical answers to the two research questions leads to conviction that LK&C can be learned and cultivated in individuals who form families and organizations in the society. If people and organizations become kinder and more compassionate, human society can accumulate more altruistic capital across space and over time, which is needed for successful implementation of strong sustainability principles.

The paper is structured as follows. Section 7.2 presents the background on why altruistic preferences and behaviors are needed for the implementation of strong sustainability principles to tackle the existing global environmental crisis. Section 7.3 introduces the framework of mental cognition to provide descriptive analysis of LK&C in Sect. 7.4. Section 7.5 provides our logical arguments to answer the two research questions. Section 7.6 provides some policy implications and concludes our paper.

7.2 Cultivation of LK&C in Individuals: A Societal Solution to Global Ecological Crisis

In this section we provide our three links in our narrative that explains why cultivation of LK&C is a necessary precondition for human society to tackle the global ecological crisis. First, strong sustainability principles must be applied in economic valuation frameworks to tackle ecological crisis. Second, operationalizing these strong sustainability principles requires more altruistic preferences and behaviors from economic agents. Third, LK&C are mental motivators for altruistic preferences and behaviors in individuals. Hence, developing LK&C is a solution to ecological crisis.

7.3 Contemporary Economic Analysis Threatens Strong Sustainability Principles

Natural resources are needed in the production and consumption of goods and services. In allocating these resources, an economic approach assumes that human and reproducible capital can substitute them, which leads to existing global ecological crisis (Stern, 1997; Cohen et al., 2019). This is a typical failure due to applying the weak sustainability principles in policies. Therefore, public policies need to uphold strong sustainability principles which reject substitutions of human and reproducible capitals for natural capital and require the preservation of all forms of natural capital (Ayres et al., 2001).

7.4 Upholding Strong Sustainability Principles Is Difficult in Public Policy Evaluation

Implementing strong sustainability principles in climate policies at state, national, or international scales is very difficult, because of few reasons as discussed in the literature (Ekins et al., 2003; DesRoches, 2019; Laitos & Okulski, 2017). First, implementing those policies is costly and has negative impacts on majority of economic agents including firms, banks, investors, families and individuals. Second, the current generation bears those costs and loss while future generations benefit from outcomes. Third, these policies have stronger impacts on poorer and more disadvantaged population groups who have less financial capacity to pay or bear the loss.

In most democratic societies, these policies need to pass through public debates which require economic valuation. However, most contemporary policy analysis frameworks including social cost and benefit analysis (CBA) could threaten strong sustainability principles. Consider two simple cases. In the first case, CBA is used to appraise projects which directly or indirectly reduce the stock of natural resource and the quality of environment. For instance, a mining project releases polluting compounds or wastes to already stressed underground water systems or coral systems in the ocean. As the natural capital has been under existing stress, the strong sustainability principle suggests no further pollution. Hence, the project should be abandoned. However, if CBA yields positive net discounted benefits (NDBs = total discounted benefits – total discounted costs), it recommends the implementation of this project. In another case, strong sustainability principle would recommend the implementation of those projects which improve the stock of critical natural capital. Examples include projects that aim to reduce carbon emission to tackle climate change. However, for these projects, CBA may yield negative NDBs, which downplay the worthiness of these projects.

Modifications are proposed to address CBA limitations. Several practical modifications include imposing higher monetary values for environmental resources, larger weights in valuating impacts of the future generations and disadvantaged population groups, and using extremely low discount rates for future generations in the flexible discounting schemes (Tol, 2001; Cai et al., 2015; Dennig, 2017).

7.5 Extremely Low Discounting Rates Are Needed

Using lower discount rates leads to a recommendation of stronger policies to combat climate change and protect critical natural environments. Guest (2010) captures this in Fig. 7.1 which displays well-being per capita per period over a 200-year time horizon (upper diagram) and cumulative well-being under the two paths using two discount rates (lower diagrams). Implementing policies, the solid green path shows marginally lower well-being for several earlier decades but remarkably higher well-being in latter decades. Using a higher discount rate (3%) yields a significant lower accumulative well-being.

Literature also argues for more flexible discounting schemes. Chichilnisky (1997) proposes to discount utilities at a declining rate approaching zero asymptotically. Defrancesco et al. (2014) proposes a multi-rate discounting scheme where each damage component is discounted at a component-specific constant rate. Galperti and Strulovici (2017) show that hyperbolic discounting regime would allow for pure and direct altruism: pure altruism concerns the total utility (rather than the mere consumption utility) of future generations and direct altruism directly incorporates the utility of all future generations.

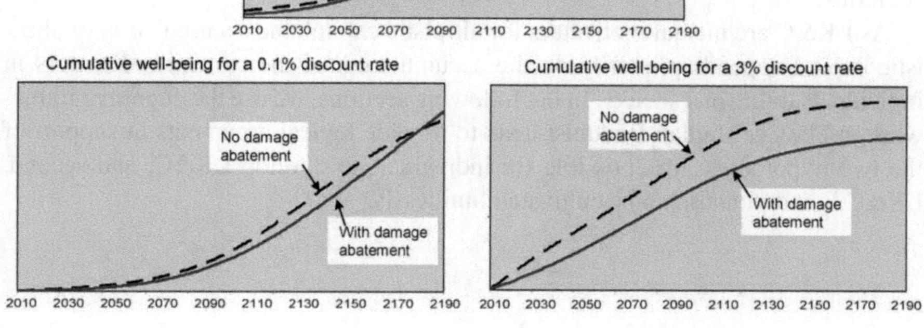

Fig. 7.1 Higher discount rate harms strong sustainability principles. (Adapted from Guest, 2010)

7.6 Altruistic Preferences and Behaviors Are Needed

Institutionalizing those modifications requires more altruistic preferences and behaviors of key economic agents. This is the implication drawn from a large body of theoretical, empirical, and experimental studies which show that altruistic individuals are found to show greater willingness to pay for pro-environmental policies and place greater importance on the benefits and loss of other people (Cardia & Michel, 2004; Ojea & Loureiro, 2007; Asensio & Delmas, 2015; Nguyen & Robinson, 2015). Similarly, economic modeling shows that altruistic preferences is consistent with the hyperbolic discounting scheme (Ekeland et al., 2015; Cao & Li, 2022). Moreover, literature classifies preferences and behavior that benefits others and future generations as pure, direct, and strong forms of altruism (Stephan & Muller-Fürstenberger, 1998; Popp, 2001; Saez-Marti & Weibull, 2005; Galperti & Strulovici, 2017). Hence, accumulation of altruistic capital in terms of altruistic preferences and behaviors in human societies is needed to deal with intergenerational issues such as sustainable development, poverty, and climate change (Ashraf & Bandiera, 2017).

7.7 LK&C Lead to Altruistic Preferences and Behaviors

LK&C are mental motivators for altruistic preferences and behaviors (Walsh-Frank, 1996; Simpson & Willer, 2015; Gilbert, 2019). Empirically, literature shows a strong connection between LK&C and pro-social and pro-environmental behaviors of individual consumers, entrepreneurs, investors, bankers and leaders (Miller et al., 2012; van Dierendonck & Patterson, 2014; Pfattheicher et al., 2016; Engel et al., 2020; Panda et al., 2020; Le Grand et al., 2021; Koller et al., 2022). Similarly, there is growing evidence of the positive impacts of Buddhism-based practices in cultivating favorable attitudes and behaviors toward sustainability in both individuals and organizations (Tsuwan, 2008; Vu, 2018; Khetjoi et al., 2020; Song, 2020; Quaglia et al., 2021). Ongoing cultivation of LK&C in individuals is needed to improve altruistic capital among societies.

As LK&C are mental motivators for altruistic capital, the accumulation of altruistic capital depends primarily on the accumulation of the two mental factors in individuals living in a society. In the following sections, we use the cognitive framework in classical Indian Buddhist texts to provide logical arguments in support of the two hypotheses: it is possible for individuals to develop LK&C; and second, LK&C in individuals can be cultivated limitlessly.

7.8 Consciousness in the Cognitive Framework

7.8.1 Definition of Consciousness

In Buddhist philosophy, consciousness[1] is defined as "that which is clear and knowing" (Rinbochay & Napper, 1981; Tsering, 2006). The "knowing/cognition" aspect refers to the ability of consciousness to know its object. There are two common ways to understand the term "clear" in this definition. First, the "clear" aspect can refer to the nonmaterial, space-like nature of consciousness which is completely devoid of color, shape, or material dimension (Rabten, 1981). It is certain that consciousness is not physical matter (Rinbochay & Napper, 1981; Tsering, 2006; Lama & Chodron, 2017; Dunne et al., 2020). Second, clarity also refers to the object's appearing "clearly" to the mind (Gyatso & DeCharms, 1997). Lama and Chodron (2017) explain that clarity in a way that "when the mind meets with certain conditions it is able to reflect objects, like a clear mirror" (p. 46). Dunne et al. (2020) put that "just as reflections appear in a mirror, any internal or external object whatsoever – good or bad, pleasant or unpleasant-can appear in consciousness, so consciousness is luminous in that it illuminates objects" (p. 41).

7.8.2 Divisions of Consciousness

Consciousness can be divided into six types: five types of sensory consciousness (i.e., eye, ear, nose, tongue, and body consciousness) and mental consciousness (Gyel-tsen, 2003; Tsering, 2006). The five sensory consciousnesses are those that have arisen in dependence on their five respective sense faculties which are their unique empowering conditions. Examples of eye consciousness are eye awareness apprehending the color red and the round shape of a red traffic light and eye consciousness that has arisen in dependence on the eye sense faculty which is a subtle physical component contained within the eye. All five types of sensory consciousness can perceive their object. For most ordinary people, sensory consciousness do not interpret, label, or describe objects (Tsering, 2006). Mental consciousness, on the other hand, in most people, apprehends the object via the medium of a generic image (also called mental images).[2] Mental consciousness relies on sensory

[1] Consciousness, awareness, and knower are synonymous in Buddhist textbooks on mind and mental factors (Rinbochay & Napper, 1981). Occasionally mind (in singular form) is used to refer to conscious experience or mental experience, while minds (in plural form) refer to the main mind (Tsering, 2006). In addition, Dunne et al. (2020) used mind, mentality, and consciousness being synonymous.

[2] In Buddhist texts on consciousness, mental consciousness can be conceptual or perceptual. Perceptual mind knows the object without depending on the generic image, while conceptual mind knows the object through the generic image (Gyel-tsen, 2003; Tsering, 2006).

consciousness to know the external objects and is able to interpret, label, describe, reason, and do many more cognitive functions.

In the Buddhist presentation of cognition, both mental consciousness and sense consciousness can be divided into their respective main minds and mental factors (Tsering, 2006). For instance, eye consciousness is one of the five types of sense consciousness, and within eye consciousness, there are a main mind and a set of mental factors in the retinue of this main mind. For mental consciousness, there is a mental main mind and many mental factors in the retinue of the mental main mind. The key difference between the main mind and the mental factors rests in the way they know an object. The main mind and its mental factors can be distinguished in the following way. The main mind knows the mere entity of an object while the mental factors know the various attributes of the same object (Gyel-tsen, 2003). In fact, Tsering (2006) defines a mental factor as the aspect of the mind that apprehends a particular quality of an object and that mental factors are active while the main mind is passive and works as a background. There are many different types of mental factors, and these mental factors can be distinguished in terms of the specific functions that each of them performs.

Since the mind's full engagement with an object occurs primarily by means of the mental factors' cognition of the object's attributes, the mental factors play an important role in the functioning of consciousness. Tsering (2006) gives an analogy of the hand of which the palm is the main mind while the fingers are the mental factors. In this metaphor, it is fingers that cause the hand to function. In other words, the mental factors influence the main mind and without mental factors the mind does not know much about the object.

With respect to mental consciousness, there are a variety of mental factors which can accompany its main mind. These mental factors can be differentiated by their specific functions. Let us illustrate specific functions of several mental factors. Feeling is one type of mental factor that has the characteristic of experience. This means that if the mental factor feeling is not present, neither the main mind nor any other mental factors would be able to experience their object in any of the three ways: pleasant, unpleasant, or neutral (Gyel-tsen, 2003; Dunne et al., 2020). Anger is another mental factor which apprehends an object being unattractive,[3] exaggerates that unattractiveness, and has ill-will toward the object. Attachment is a mental factor which apprehends an object as attractive, exaggerates that attractiveness, and strives for or clings to the object. It is also important to note that for ordinary beings, many mental factors can arise together (at the same time) and at that moment, they have the same focal object. For instance, the mental factors of feeling, loving-kindness, compassion, and wisdom can occur at the same time, while each of these has their own specific functions apprehending specific attributes of the focal object.

Importantly, mental factors are grouped into virtuous, non-virtuous, and neutral divisions. According to Dunne et al. (2020), the overall distinction between the first two divisions is based largely on the immediate and long-term effects that they

[3] "Attractive" and "unattractive" do not necessarily relate to mere physical appeal of the object.

induce. In immediate terms, the virtuous factor functions in a way that prevents negative or destructive behavior, while non-virtuous mental factors function to induce unpleasant mental states and induce negative, destructive behaviors. In the long term, the virtuous factor produces happiness, while non-virtuous mental factors result in suffering.

7.9 LK&C: Characteristics, Variations, and Inhibitors

7.9.1 Characteristics of LK&C

In most Buddhist traditions, loving-kindness is often understood as the wish for sentient beings to be happy and compassion as the wish for sentient beings to be free from suffering.[4] Among many mental factors discussed in Buddhist texts, both compassion and love could be argued to be subsumed under the mental factor of "non-hatred." According to Gyel-tsen (2003), non-hatred is "a lack of malice with regard to sentient beings, suffering and the sources of suffering. It has the function of acting as a support for non-engagement in misconduct" (p.37).

There is a distinction between loving-kindness and compassion. Dunne et al. (2020) describe that "compassion arises from observing sentient beings to be suffering and has the aspect of wishing them to be free of that. Love, on the other hand, arises upon observing sentient beings from the perspective of their well-being and has the aspect of wishing them to be happy" (p.126). Clearly, in both cases LK&C focus on sentient beings; however, they differ in their respective subjective aspects. In general, a state of happiness would be a state that is free of all types of suffering. Hence, LK&C are strongly related to each other. In fact, most Buddhist teachings promote the cultivation of these two mental states together (Lama & Chodron, 2017, 2020; Dunne et al., 2020).

7.9.2 Variations of LK&C

In this section, we attempt to summarize variations of LK&C according to their focal objects and the wisdom influencing them. In ordinary people, there are many variations of LK&C, for example, a mother's LK&C toward their infant child or the LK&C that ordinary people extend toward their relatives, friends, and animals they care about or who help them (Dunne et al., 2020; Lama & Chodron, 2020). Ordinary people can also extend LK&C toward people and animals that they do not know.

[4] Literature also discusses the type of LK&C toward oneself. It is logical that without wishing oneself to be happy and free from suffering, it is impossible for oneself to engage in activities that lead to happiness. However, in most Buddhist texts, LK&C are discussed mainly in relation to beings other than oneself.

However, in general, their LK&C are confined to wishing these beings to have good health, wealth, good job, families, and be free from physical and mental pains in this life (Lama & Chodron, 2020).

In fact, for ordinary people, when mixed with attachment to the focal object (as one's relative or friend), LK&C are biased and limited in their scope. If the associated attachment diminishes, likewise the LK&C can be easily lost. As Lama and Hopkins (1988) comment: "there is compassion that is mixed with desire; this is the kind that we usually have, for when some little thing goes wrong, we immediately get angry" (p. 168). Additionally, when mixed with attachment, there is little opportunity for the development of LK&C toward those individuals one might categorize as "enemies." However, when based on an unbiased and even attitude toward beings, regardless of whether they be considered friend, enemy, or neither of those two, and a wisdom that accurately assesses the situation of others in lacking happiness and possessing suffering, LK&C are considered genuine (Lama, 2000). Not only do such unbiased LK&C not diminish if afflictions are not present, but their mode of development itself serves to reduce afflictions themselves through an accurate assessment of the object of their focus.

In another way, one can also relate LK&C in the "four immeasurables" taught in various Buddhist traditions: equanimity, love, compassion, and empathic joy as detailed in Lama and Chodron (2020). Particularly, in the Mahayana traditions, with immeasurable loving-kindness, the wish that sentient beings have happiness intensifies and becomes an aspiration which then grows into a resolution to get involved and take responsibility to make the aspiration a reality (Lama & Chodron, 2020). Similarly, with immeasurable compassion, the wish that sentient beings be free from suffering and its causes has stronger intensities.

On the higher level, literature discusses the great compassion that the practitioners of Mahayana traditions are striving for through the stages of development toward the higher quality of the great compassion of Bodhisattvas.[5] Tsongkhapa and Jinpa (2021) provide excellent explanations of three types of Bodhisattvas' great compassion: compassion observing sentient beings, compassion observing phenomena, and compassion observing the unapprehend able. All types of great compassion have same focal objects: all sentient beings that are submerged in the three kinds of sufferings: suffering of physical and mental pains, sufferings of change, and pervasive suffering of conditioning. All have the same subjective aspect of wholehearted determination to protect all sentient beings from *samsaric* suffering. Importantly, three types of great compassion are differentiated by the quality of wisdom that influences compassion. The great compassion observing sentient beings is not explicitly qualified by the wisdom realizing subtle impermanence and emptiness. The last two types of great compassion are compassion explicitly

[5] Bodhisattvas refer to beings who have attained the uncontrived mind of enlightenment. Their mind of enlightenment is uncontrived in the sense that Bodhisattvas can manifest their mind of enlightenment effortlessly. The mind of enlightenment refers to the mind that strives for the full enlightenment for the sake of all sentient beings. See Tsering (2008) and Lama and Chodron (2020) for more discussions on the mind of enlightenment.

qualified by the force of wisdom realizing subtle impermanence and emptiness, respectively.

7.9.3 Inhibitors to LK&C

Contemporary Western psychology literature discusses fears, blocks, and resistances as inhibitors that prevent LK&C (Gilbert & Mascaro, 2017; Kirby et al., 2019). In a similar vein, one can view mental afflictions as mental inhibitors of virtuous mental afflictions. In a classic Buddhist text, non-virtuous mental factors are grouped into 6 root afflictions and 20 secondary afflictions (Gyel-tsen, 2003).[6] From the cognitive framework, when focused on the same object, the manifest generation of afflictions such as anger and hatred will prevent the generation of opposite mental states LK&C, when focused on the same object. This is due to the incompatibility of their subjective aspects. The subjective aspect of anger is wishing to harm sentient beings, and this aspect is opposite to wishing to benefit them (as the aspect of loving-kindness) and wishing them to be free from suffering (as the aspect of compassion). In addition, Lama and Chodron (2019) note that:

> We cannot simultaneously experience two manifest mental states that are contradictory – we cannot be angry and loving at exactly the same moment. When we are loving, anger is not manifest in our minds, but we haven't eliminated anger from our mindstreams either. The seed of anger remains on our mind streams when love is manifest, and it connects one instance of anger to the next. (p.128)

7.10 Definite Potential for Loving Kindness and Compassion[7]

As discussed earlier, solving strong sustainability issues within contemporary economic analysis framework requires agents to have preferences for more altruistic preferences and behaviors. From the Buddhist perspective, LK&C are mental motivators for altruistic preferences and behaviors. In this section, we attempt to analyze the two following hypotheses:

Hypothesis 1: It is possible for individuals to generate LK&C.
Hypothesis 2: Individuals can cultivate LK&C limitlessly.

[6] This list of mental afflictions is not exhaustive. The Shakyamuni Buddha spoke of 84,000 mental afflictions (Lama & Chodron, 2019).

[7] There is extensive literature on methods to cultivate LK&C from both Buddhist practices, which lays foundations for contemporary compassion trainings and therapies in Western psychology (Gilbert, 2019; Quaglia et al., 2021).

We will use three logical arguments to support these hypotheses. First, it is possible for individuals to generate LK&C because they have the mental main mind and because the mental inhibitors to LK&C can be reduced and eliminated completely. Its logic is as follows: as long as the mental main mind exists, its capacity for the generation of LK & C remains. Without mental main mind, it is impossible for other mental consciousnesses such as the mental factors of LK&C to exist; these mental factors are generated in dependence upon potencies that exist within the main mind. Second, mental inhibitors can be reduced and eliminated completely because they are (1) adventitious and (2) operate on the basis of wrong cognitions. Third, LK&C can be developed limitlessly because they have (1) a stable basis and (2) operate based on valid cognitions. This third argument is also applicable in the context of wisdom as the antidote to ignorance. We discuss these in turn.

7.10.1 The First Logical Argument

The reason in the first logical proof is twofold: (1) individuals have the mental main mind and (2) mental inhibitors to LK&C can be reduced and eliminated completely. For the first part, undoubtedly, individuals have the mental main mind, and mental main mind is one of the necessary factors for the generation of all mental factors associated with it. As this mental main mind does not cease to exist, the potential for mental factors to exist and arise remains in individuals. The reasoning for the second part rests on the second logical proof.

7.10.2 The Second Logical Argument

As Lama and Chodron (2019) postulate, "some people assert that afflictions are an inherent part of human nature and, as such, are hardwired in our nervous system or genes. Although we may be able to modify their effects, we can never be free of them" (p. 106). Such assertions can be proven wrong through the following two points.

7.11 Mental Afflictions Including Anger and Hatred Are Adventitious

Dharmakirti posits that "The nature of mind is clear light. The defilements are adventitious" (p. 161). Mental afflictions are adventitious in the sense that they are not the very nature of the mind and they do not penetrate into the basic nature of the mind (Lama & Chodron, 2019).[8] If mental afflictions had ever penetrated the very nature of the mind, they would always be present rather than temporary. One can easily cognize that anger is temporary in the sense that it is sometimes present and sometimes not. One can also understand a specific meaning of "mental afflictions are adventitious" in the sense that when afflictions meet their powerful antidotes, these afflictions cannot remain.

7.12 Mental Afflictions Operate on the Basis of Wrong Cognition

Lama (2000) defines mental afflictions as "a state which causes disturbance within one's mind" (p. 28), which then lead to faulty physical and verbal behavior. These mental afflictions arise in dependence upon a faulty cognitive process. Lama and Chodron (2019) present an elaborative discussion by Tsongkhapa that:

> When the view of a personal identity [which is ignorance] apprehends the self, discrimination arises between self and other. Once you have made that distinction, you become attached to what is associated with yourself and hostile toward that which pertains to others. As you observe the self, your mind becomes inflated [with arrogance]. You develop a belief that this very self is either eternal or subject to annihilation [view of extremes]. (p. 105)

From the cognitive framework, there is a causal mental process that regulates how non-virtuous afflictions operate on the basis of wrong cognitions. At the root of this process, there is innate fundamental ignorance that causes incorrect mentation which in turn causes the generation of non-virtuous afflictions.[9] Tsongkhapa and Jinpa (2021) present Candrakirti's discussions as follows:

> "Delusion" refers to ignorance, for it apprehends things as real; as such, it operates by way of exaggeration superimposing truly existent natures upon things. Attachment and so on also operate by thoroughly attributing qualities of attractiveness or unattractiveness upon the nature of things imputed by delusion, and as such, they do not operate independently of delusion. Furthermore, they are contingent on delusion in that delusion is the primary factor. (p. 241)

[8] The statement "the nature of mind is clear light" does not refer to the "clear" and "knowing" aspects of consciousness as discussed earlier. The analogy is that muds (in muddy water) do not come into the nature of water, and when all muds are removed from the water, the water becomes clean.

[9] The Middle Way Consequence school (Prasangika School) considers two types of afflictive obscuration: active forms of afflictions and the seeds of these afflictions. The seeds themselves are not afflictions. The seeds are substantial causes for their respective resultant afflictions, meaning that the seeds have potency to manifest in the active forms of afflictions. For instance, in the context of anger, when anger in an ordinary being subdues, the seed of anger remains in his or her mind stream, and it connects the prior and later instances of anger. See more discussions in Chap. 5 of Lama and Chodron (2019).

While attachment (respectively anger) sees the object as being attractive (unattractive), Tsongkhapa and Jinpa (2021) clarified that "attractive and unattractive qualities are superimposed by incorrect mentation, which is in fact a cause of attachment and aversion" (p. 241). In fact, incorrect mentation causes exaggerated attractiveness (unattractiveness), and this exaggerated level of attractiveness (unattractiveness) appears to attachment (anger). Due to its nature of clarity and cognition, attachment (anger) simply takes for granted that the object has such exaggerated attractiveness appearing to it. Hence, the Candrakirti commentary mentioned above should be understood as meaning "attachment and aversion operate on the basis of superimposing intrinsically existing qualities of attractiveness and unattractiveness" (Tsongkhapa & Jinpa, 2021, p.241). Furthermore, Tsongkhapa and Hopkins (2014) present a series of arguments clarifying what type of ignorance can be identified as the root of all mental afflictions. The authors conclude that:

> Therefore, both [apprehensions of the true existence of persons and apprehensions of the true existence of phenomena] should be taken to be ignorance. All other afflictive emotions—innate and artificial—operate within apprehending individual features of just that object on which the innate ignorance described above has superimposed [a sense of inherent existence]. Therefore, it is said that just as the other four sense powers – eyes and so forth—abide in dependence on the body sense power and are not located in a place under their own power other than [where the body sense power is], so even all other afflictive emotions operate in dependence on innate ignorance, due to which bewilderment is chief. (p. 50–51)

As all non-virtuous afflictions are adventitious and operate on the basis of wrong cognitions rooted in fundamental ignorance, they can be decreased and completely removed in individuals. This is because when fundamental ignorance and wrong cognitions are reduced and removed, these mental afflictions will reduce and disappear completely. However, this line of reasoning poses another question: can ignorance and wrong cognition can be reduced and removed completely? One simple answer is that through education, individuals can learn, develop, and improve their knowledge, thereby reducing their "not knowing." However, to prove that ignorance and wrong cognition can be eradicated completely, one needs to rely on the logical argument supporting the view that wisdom and valid cognitions can be increased limitlessly.

7.13 The Third Logical Argument

LK&C are similar to their inhibitors in that they all are adventitious. However, unlike anger and attachment, LK&C do not operate with the support of wrong cognitions. In fact Lama and Hopkins (1988) postulate this position for all mental afflictions and said "Love and compassion have the support of valid cognition; their production does not need assistance from ignorance that misconceives objects to be inherently existent…Non-virtuous attitudes, such as hatred, pride and so forth—no matter how strong they are—are generated *only* with the assistance and support of a conception of inherent existence. Therefore, without such misconception, there is no way that desire and hatred can operate, but consciousnesses that are the opposites of desire and hatred can operate even when there is no misconception of inherent existence" (p.102).

One might argue that some forms of LK&C are based on wrong cognitions. For instance, compassion focused on people who are not suffering physical or mental pains is not based on a valid cognition that identifies them as suffering because these people are not suffering. The fundamental Buddhist teaching on the truth of suffering counteracts this argument. In fact, Lama and Chodron (2020) state that:

> Our compassion is also limited in that it arises for those experiencing the duḥkha of pain – what all sentient beings consider as undesirable – but not for those who are young, healthy, rich, successful, talented, powerful, artistic, athletic, or attractive. Thinking that they are not suffering, we don't feel compassion for them. We forget that they are imprisoned in saṃsāra owing to afflictions and polluted karma, and we neglect to consider that they're experiencing the duḥkha of change and the pervasive duḥkha of conditioning.[10] (p.171)

Prior to proving that LK&C in individuals can develop limitlessly, one needs to understand the word "limitlessly." As mentioned above, LK&C are supported by valid cognitions and, since their support is the clear cognizing mind, can be familiarized with to the point at which the effort and exertion required at first for their generation are no longer required in order for them to arise. At this point of familiarity, LK&C arise naturally at all times, whether one is eating, talking, walking about, or sitting. If this level of familiarity is gained, then through habituating further with these states of mind over many lifetimes, through eliminating all faults whatsoever from the mind, in the end they become omniscience. Thus, the logical proof for the second hypothesis rests on the following components: (1) LK&C have a stable basis and (2) LK&C operate on the basis of valid cognitions.

The stable basis of LK&C is the clarity and cognition nature of the consciousness. If the basis for these consciousnesses were a physical form, then there would be a limit to the extent to which this physical form can develop through training. Lama (2006) provides a simple example of physical training in broad-jumping in which the basis of this training process is the gross physical body, and thus there is a limit to how much a person can jump. Further, each jump requires a new exertion; it is not possible to reach a point at which, due to previous familiarization with jumping, effort and exertion are no longer required for jumping. This is due to the limits of the basis (the body). However, for LK&C trainings, the basis of this mental development process is the mind; hence, Lama and Hopkins (1988) say that "due to this, once one has trained in them, one does not need to rely on repeated exertion to regenerate them and thus they can become stronger and stronger. Consequently, they can be developed to a limitless degree. Because of these reasons – that compassion, wisdom and so forth are qualities that depend on the mind, and the mind is stable and continuous – they can be developed to a limitless degree" (p.103).

[10] Saṃsāra refers to cyclic existence and duḥkha sufferings. Further explanation of these other types of suffering can be found in Lama and Chodron (2019).

This line of reasoning can be traced back to the reason that Dharmakirti provided in his treatise on valid cognition in proving that individuals can develop LK&C to the "uncontrived" level in which LK&C can be become effortless. Jackson and Dar-ma-rin-chen (1993) describe that "At first, taking all sentient beings as one's intentional object, one generates through effort a forceful compassion that desires to free them from suffering. When one has generated that [compassion] by great effort, each effort is able to generate many homogenes of compassion. Through such accustomation, one comes to have a nature such that great compassion naturally enters [the mindstream] without depending on such effort, but [such accustomation] is not independent of effort from the first" (p. 318).

Here a doubt may arise: if LK&C have a basis that is the mere clear and knowing mind, then the afflictions, such as anger and attachment, must also have such a basis and therefore may be developed limitlessly. Although afflictions can indeed be familiarized with to a great degree, as mentioned previously, such afflictions do not possess the support of valid cognitions but rather are based on wrong cognitions. Hence, when the antidotes to those wrong cognitions are applied, they will reduce and eventually may be completely eliminated (Lama & Hopkins, 1988). Although when LK&C are weak and not yet familiarized with, they may be harmed by their inhibitors; once familiarization has been gained to a great extent, this can no longer occur.

One can approach understanding this through examining the process of developing the mind of enlightenment as instructed in two primary methods of meditation in Mahayana traditions. The first method has seven steps in total, in which prior to developing LK&C, one needs to develop three types of consciousness: recognizing all sentient beings as mothers, becoming mindful of their kindness, and developing an intention to repay their kindness. So, in this process, LK&C are the results of those cognitions which are valid. The second method has the first step of cultivating the awareness of equalizing self and others. For this type of awareness, one needs to recognize that all sentient beings—oneself and all others—are equal in many senses.

It is not possible to logically argue against the position that learning increases knowledge commensurate with the degree of familiarization and habituation with that learning. Similarly, it is reasonable to accept that LK&C can be increased and developed more and more through the process of familiarization and habituation. In fact, Buddhist texts present different techniques and instructions for this very purpose. We would like to conclude this section by quoting the following discussion by Lama (2006) on the possibility that love will eventually overcome hatred because the former is valid while the latter is a wrong consciousness:

> What are the circumstances that serve as a basis for generating hatred? It is generated because we superimpose upon phenomena an unattractiveness or badness that exceeds what is actually there. With this as a basis we get angry at what prevents our desires. Hence, the foundation of a mind of hatred is not valid. However, a mind of love does have a valid foundation. When, over a long period of time, an attitude that has a valid foundation competes with an attitude that does not, the one with the valid foundation will win.
>
> Therefore, if you familiarize steadily over a long period of time with good attitudes that have a valid foundation, bad attitudes that do not will gradually diminish. (p. 26)

7.14 Conclusion

Literature shows that LK&C are mental motivators for altruistic preferences and behaviors (Walsh-Frank, 1996; Simpson & Willer, 2015; Gilbert, 2019). In addition, empirical studies establish a strong connection between LK&C and pro-social and pro-environmental behaviors of individual consumers, entrepreneurs, investors, bankers, and leaders (Miller et al., 2012; van Dierendonck & Patterson, 2014; Pfattheicher et al., 2016; Engel et al., 2020; Panda et al., 2020; Le Grand et al., 2021; Koller et al., 2022). As altruistic preferences and behaviors in individuals and organizations form altruistic capital, ongoing cultivation of LK&C is needed to improve altruistic capital among societies. This chapter extends these strands of literature by providing a narrative showing that altruistic capital is also crucial for successful implementation of strong actions, strong policies, and the upholding of strong sustainability principles to deal with the existing global ecological crisis.

Based on the fact that altruistic preferences and behaviors in individuals are motivated by LK&C, we provide our descriptive analysis of these two mental states from the cognitive perspective which is based primarily on the classical Buddhist texts on consciousness. Instead of focusing on how individuals cultivate LK&C as well as their benefits as discussed elsewhere, our analysis provides logical supports for the two hypotheses: (1) it is possible for individuals to generate LK&C and (2) individuals can cultivate LK&C limitlessly. We believe that developing conviction in these two possibilities would allow scholarly debate to focus on how public policies should be designed to best to allocate limited resources toward enhancing LK&C in individuals. While discussions in this article are drawn from Buddhist literature, implications on how best to train individuals should consider diversity of their religious, socioeconomic, and cultural backgrounds.

From the training perspective, this paper also highlights the importance of valid cognitions that refute the mental inhibitors against LK&C as well as the valid awareness that lay the foundations for the limitless enhancement of LK&C. Therefore, trainings should not only focus on either cultivating LK&C or removing inhibitors themselves but rather should be combined with developing knowledge and wisdom.

References

Asensio, O. I., & Delmas, M. A. (2015). Nonprice incentives and energy conservation. *Proceedings of the National Academy of Sciences, 112*, E510.

Ashraf, N., & Bandiera, O. (2017). Altruistic capital. *American Economic Review, 107*, 70.

Augustine, P., & Wayne, M. (2019). Understanding the phenomenon: A comparative study of compassion of the West and karuna of the East. *Asian Philosophy, 29*, 1.

Ayres, R., Van den Berrgh, J., & Gowdy, J. J. E. E. (2001). Strong versus weak sustainability: Economics, natural sciences, and consilience. *Environmental Ethics, 23*, 155.

Cai, Y., Judd, K. L., Lenton, T. M., Lontzek, T. S., & Narita, D. (2015). Environmental tipping points significantly affect the cost and benefit assessment of climate policies. *Proceedings of the National Academy of Sciences, 112*, 4606.

Cao, J., & Li, M. (2022). Hyperbolic discounting in an intergenerational model with altruistic parents. *Journal of Population Economics, 35*, 989.

Cardia, E., & Michel, P. (2004). Altruism, intergenerational transfers of time and bequests. *Journal of Economic Dynamics Control, 28*, 1681.

Chichilnisky, G. (1997). What is sustainable development? *Land Economics, 73*, 467.

Cohen, F., Hepburn, C. J., & Teytelboym, A. (2019). Is natural capital really substitutable? *Annual Review of Environment and Resources, 44*, 425.

Conway, J. (2001). A Buddhist critique of Nussbaum's account of compassion. *Philosophy in the Contemporary World, 8*, 7.

Costanza, R. (2020). Valuing natural capital and ecosystem services toward the goals of efficiency, fairness, and sustainability. *Ecosystem Services, 43*, 101096.

Daily, G. C., Söderqvist, T., Aniyar, S., Arrow, K., Dasgupta, P., Ehrlich, P. R., Folke, C., Jansson, A., Jansson, B.-O., Kautsky, N., Levin, S., Lubchenco, J., Mäler, K.-G., Simpson, D., Starrett, D., Tilman, D., & Walker, B. (2000). The value of nature and the nature of value. *Science, 289*, 395.

Defrancesco, E., Gatto, P., & Rosato, P. (2014). A 'component-based' approach to discounting for natural resource damage assessment. *Ecological Economics, 99*, 1.

Dennig, F. (2017). Climate change and the re-evaluation of cost-benefit analysis. *Climatic Change, 151*, 43.

DesRoches, C. T. (2019). Some truths don't matter: The case of strong sustainability. *Ethics, Policy & Environment, 22*, 184.

Dunne, J. D., Jinpa, T., & Rochard, D. (2020). *Science and philosophy in the Indian Buddhist classics, vol. 2: The mind.* Wisdom Publications.

Ekeland, I., Karp, L., & Sumaila, R. (2015). Equilibrium resource management with altruistic overlapping generations. *Journal of Environmental Economics and Management, 70*, 1.

Ekins, P., Simon, S., Deutsch, L., Folke, C., & De Groot, R. (2003). A framework for the practical application of the concepts of critical natural capital and strong sustainability. *Ecological Economics, 44*, 165.

Engel, Y., Ramesh, A., & Steiner, N. (2020). Powered by compassion: The effect of loving-kindness meditation on entrepreneurs' sustainable decision-making. *Journal of Business Venturing, 35*, 105986.

Eyring, V., Gillett, N. P., Achuta Rao, K. M., Barimalala, R., Parrillo, M. B., Bellouin, N., Cassou, C., Durack, P. J., Kosaka, Y., McGregor, S., Min, S., Morgenstern, O., & Sun, Y. (2021). Human influence on the climate system. In V. Masson-Delmotte, P. Zhai, A. Pirani, S. L. Connors, C. Péan, S. Berger, N. Caud, Y. Chen, L. Goldfarb, M. I. Gomis, M. Huang, K. Leitzell, E. Lonnoy, J. B. R. Matthews, T. K. Maycock, T. Waterfield, O. Yelekçi, R. Yu, & B. Zhou (Eds.), *Climate change 2021: The physical science basis. Contribution of Working Group I to the Sixth Assessment Report of the Intergovernmental Panel on Climate Change*. Cambridge University Press, in press.

Galperti, S., & Strulovici, B. (2017). A theory of intergenerational altruism. *Econometrica, 85*, 1175.

Gilbert, P. (2019). Explorations into the nature and function of compassion. *Current Opinion in Psychology, 28*, 108.

Gilbert, P., & Mascaro, J. (2017). Compassion: Fears, blocks, and resistances: An evolutionary investigation. In *The Oxford handbook of compassion science* (Vol. 399). Oxford University Press.

Guest, R. (2010). The economics of sustainability in the context of climate change: An overview. *Journal of World Business, 45*, 326.

Gyatso, L., & DeCharms, R. C. (1997). A discussion with the venerable Logsang Gyatso: Mind and mental factors. In R. C. DeCharms (Ed.), *Two views of mind: Abhidharma and brain science*. Snow Lion Publications.

Gyel-tsen, Y.-S. (2003). *A necklace for those of clear awareness clearly revealing the modes of minds and mental factors*. Toh Sze Gee & FPMT. FPMT Publication.

Jackson, R. R., & Dar-ma-rin-chen, R.-T. (1993). *Is enlightenment possible?: Dharmakirti and Rgyal Tshab Rje on knowledge, rebirth, no-self and liberation*. Snow Lion.

Kelleher, J. P. (2017). Descriptive versus prescriptive discounting in climate change policy analysis. *Georgetown Journal of Law & Public Policy, 15*, 957.

Khetjoi, S., Srisaket, D., Chompromma, P., & Khetjoi, S. (2020). Buddhism: The contribution to socio-economy sustainable development. *Journal of Buddhist Education and Research, 6*, 1.

Kirby, J. N., Day, J., & Sagar, V. (2019). The 'flow' of compassion: A meta-analysis of the fears of compassion scales and psychological functioning. *Clinical Psychology Review, 70*, 26.

Koller, S., Stephan, U., & Ahmetoglu, G. (2022). Ecological rationality and entrepreneurship: How entrepreneurs fit decision logics to decision content and structure. *Journal of Business Venturing, 37*, 106221.

Laitos, J., & Okulski, J. (2017). *Why environmental policies fail*. Cambridge University Press.

Lama, D. (2000). *The heart of the Buddha's path*. Books 66 Ltd.

Lama, D. (2006). *Kindness, clarity, and insight*. Snow Lion Publications.

Lama, D., & Chodron, T. (2017). *Approaching the Buddhist path*. Simon and Schuster.

Lama, D., & Chodron, T. (2019). *Samsara, nirvana, and Buddha nature*. Simon and Schuster.

Lama, D., & Chodron, T. (2020). *In praise of great compassion*. Simon and Schuster.

Lama, D., & Chodron, T. (2021). *Courageous compassion*. Simon and Schuster.

Lama, D., & Hopkins, J. (1988). *The Dalai Lama at Harvard: Lectures on the Buddhist path to peace*. Snow Lion Publisher.

Le Grand, J., Roberts, J., & Chandra, G. (2021). *Buying for good: Altruism, ethical consumerism and social policy*. Social Policy & Administration.

Lind, R. C. (1995). Intergenerational equity, discounting, and the role of cost-benefit analysis in evaluating global climate policy. *Energy Policy, 23*, 379.

Makransky, J. (2021) *Compassion in Buddhist psychology*. Compassion and Wisdom in Psychotherapy, edited by Christopher K.Germer and Ronald D. Siegel, Guilford Press, 2012

Martin, D., Seppala, E., Heineberg, Y., Rossomando, T., Doty, J., Zimbardo, P., Shiue, T.-T., Berger, R., & Zhou, Y. (2014). Multiple facets of compassion: The impact of social dominance orientation and economic systems justification. *Journal of Business Ethics, 129*, 237.

Miller, T. L., Grimes, M. G., McMullen, J. S., & Vogus, T. J. (2012). Venturing for others with heart and head: How compassion encourages social entrepreneurship. *The Academy of Management Review, 37*, 616.

Nguyen, T. C., & Robinson, J. (2015). Analysing motives behind willingness to pay for improving early warning services for tropical cyclones in Vietnam. *Meteorological Applications, 22*, 187.

Nunez, S., Arets, E., Alkemade, R., Verwer, C., & Leemans, R. (2019). Assessing the impacts of climate change on biodiversity: Is below 2 °C enough? *Climatic Change, 154*, 351.

Ojea, E., & Loureiro, M. L. (2007). Altruistic, egoistic and biospheric values in willingness to pay (Wtp) for wildlife. *Ecological Economics, 63*, 807.

Panda, T. K., Kumar, A., Jakhar, S., Luthra, S., Garza-Reyes, J. A., Kazancoglu, I., & Nayak, S. S. (2020). Social and environmental sustainability model on consumers' altruism, green purchase intention, green brand loyalty and evangelism. *Journal of Cleaner Production, 243*. 118575

Pearce, D., Atkinson, G., & Mourato, S. (2006). *Cost-benefit analysis and the environment: Recent developments*. Organisation for Economic Co-operation and Development.

Pelenc, J., & Ballet, J. (2015). Strong sustainability, critical natural capital and the capability approach. *Ecological Economics, 112*, 36.

Pfattheicher, S., Sassenrath, C., & Schindler, S. (2016). Feelings for the suffering of others and the environment: Compassion fosters proenvironmental tendencies. *Environment and Behavior, 48*, 929.

Popp, D. (2001). Altruism and the demand for environmental quality. *Land Economics, 77*, 339.

Quaglia, J. T., Soisson, A., & Simmer-Brown, J. (2021). Compassion for self versus other: A critical review of compassion training research. *The Journal of Positive Psychology, 16*, 675.

Rabten, G. (1981). *The mind and its functions*. Tharpa Choeling.

Rinbochay, L., & Napper, E. (1981). *Mind in Tibetan Buddhism*. Snow Lion Publications.
Saez-Marti, M., & Weibull, J. W. (2005). Discounting and altruism to future decision-makers. *Journal of Economic Theory, 122*, 254.
Simpson, B., & Willer, R. (2015). Beyond altruism: Sociological foundations of cooperation and prosocial behavior. *Annual Review of Sociology, 41*, 43.
Song, H. C. (2020). Sufficiency economy philosophy: Buddhism-based sustainability framework in Thailand. *Business Strategy and the Environment, 29*, 2995.
Stephan, G., & Muller-Fürstenberger, G. (1998). Discounting and the economic costs of altruism in greenhouse gas abatement. *Kyklos, 51*, 321.
Stern, D. (1997). The capital theory approach to sustainability: A critical appraisal. *Journal of Economic Issues, 31*, 145.
Tol, R. S. J. (2001). Equitable cost-benefit analysis of climate change policies. *Ecological Economics, 36*, 71.
Tsering, T. (2006). *Buddhist psychology: The Foundation of Buddhist thought, volume 3*. Simon and Schuster.
Tsering, T. (2008). *The awakening mind: The foundation of Buddhist thought, volume 4*. Simon and Schuster.
Tsongkhapa, J., & Hopkins, J. (2014). *Tsongkhapa's final exposition of wisdom*. Shambhala Publications.
Tsongkhapa, J., & Jinpa, T. (2021). *Illuminating the intent: An exposition of Candrakirti's entering the middle way*. Wisdom Publications.
Tsuwan, C. (2008). *Buddhist perspectives on sustainability: Towards radical transformation of self and world*. Doctoral dissertation, RMIT University, Australia.
van Dierendonck, D., & Patterson, K. (2014). Compassionate love as a cornerstone of servant leadership: An integration of previous theorizing and research. *Journal of Business Ethics, 128*, 119.
Victor, P. A. (2020). Cents and nonsense: A critical appraisal of the monetary valuation of nature. *Ecosystem Services, 42*, 101076.
Vu, M. C. (2018). Skilful means – A Buddhist approach to social responsibility. *Social Responsibility Journal, 14*, 321.
Walsh-Frank, P. (1996). Compassion: An east-west comparison. *Asian Philosophy, 6*, 5.
White, R. (2017). Compassion in philosophy and education. In P. Gibbs (Ed.), *The pedagogy of compassion at the heart of higher education* (p. 19). Springer International Publishing.
Yao, F. (2006). There are no degrees in a Bodhisattva's compassion. *Asian Philosophy, 16*, 189.

Chapter 8
Calling to the Mind: Exploring the Interlinkages of Mindfulness, Morality, and Ecological Sustainability

Mai Chi Vu and Nadia Singh

Abstract Ecosystems across the world are facing catastrophic effects due to high degree of environmental pressures coupled with lack of ecological consciousness among a large section of the society. This is partly attributed to the fact that people tend to equate their well-being with enhanced consumption, and material accumulation, and are reluctant to adopt lifestyle changes toward sustainable consumption. However, a number of recent studies demonstrate that human well-being is rooted in a complex array of psychological factors and sociological influences, rather than material wealth alone. In this chapter we contribute to these existing debates by analyzing how mindfulness can be used as a tool to promote ecological sensitivity and ethical behaviors among practitioners. Based on empirical research among Buddhist and Sikh mindfulness practitioners in India and Vietnam, the study demonstrates how ethical dimensions of mindfulness can help to motivate an orientation toward sustainability and other centeredness. These in turn pay a "double dividend" in terms of contribution to a sustainable way of life as well as a greater sense of well-being.

Keywords Mindfulness · Environmental consciousness · Loving-kindness · Well-being · Sustainability

M. C. Vu (✉) · N. Singh
Newcastle Business School, Northumbria University, Newcastle upon Tyne, UK
e-mail: mai.c.vu@northumbria.ac.uk; nadia.singh@northumbria.ac.uk

N. Singh et al. (eds.), *Faith Traditions and Sustainability*, Management, Change, Strategy and Positive Leadership, https://doi.org/10.1007/978-3-031-41245-5_8

8.1 Introduction

Ecological degradation remains a core challenge in developing countries owing to a weak legislative framework, lack of regulation, and absence of adequate penalties for polluting firms (Singh & Singh, 2019). Due to this there is lack of adequate incentives for firms to adopt robust environmental management systems and reduce the ecological footprint of their business activities (Pye, 2019). In the face of weak government responses to ecological challenges in developing countries, there are growing calls to explore the role of informal institutions such as grassroot activists, NGOs, faith leaders, and community-based organizations in compelling businesses and firms to adopt sustainable practices (Roxas, 2021). In this chapter, we examine spirituality and spiritual practices such as mindfulness as an informal institution can facilitate environmental activism toward sustainability in developing country contexts for a number of reasons.

First, in recent years both secular and faith-based organizations are exploring the influence of mindfulness as a possible solution to contemporary environmental problems (Brown & Kasser, 2005; Edwards, 2015; Ericson et al., 2014). It is contended that mindfulness practice can promote enhanced cognition of how an individuals' thoughts, feelings, morals, and experience lead them to adopt unsustainable behaviors. This further facilitates self-regulation and enhanced environmental consciousness through a sense of interconnectedness to the natural environment (Barbaro & Pickett, 2016). In 2015, a number of scholars started referring to "ecological mindfulness" as a new paradigm of sustainability (Mueller & Greenwood, 2015; Sol & Wals, 2015). Ecological mindfulness is conceptualized as a set of practices that help to alter individual capacities, behaviors, and lifestyles at the individual, as well as community level to enhance ecological sustainability (Sols & Wals, 2015). A few studies that show how "moment awareness," a key tenet of mindful practices, can support healthy lifestyles and pro-environmental attitudes (Geiger et al., 2018). Yet, ecological mindfulness has largely remained a theoretical construct. In this chapter, we analyze how mindfulness practices can promote new forms of environmental activism and create novel pathways toward sustainable development.

Second, existing scholarship has largely ignored how mindfulness practices are differentiated by cultural systems and religious traditions (Chinn, 2015). This is particularly relevant in developing country contexts. In these contexts, spiritual/religious practices remain a strong influence on sustainability activism (Singh & Singh, 2019). Through a cross-cultural study of Vietnam and India, this chapter explores how mindfulness practitioners are contributing to local and national projects in both the studied contexts. These exploratory case studies will provide a template to analyze these mindful-based influences on issues of environmental protection in other contexts as well. The two case studies were chosen because both Vietnam and India suffer from high levels of environmental degradation, characterized by industrial wastewater, biodiversity losses, and high levels of toxic air pollution (Tran et al., 2020). According to the Global Environmental Performance Index,

2020, India ranked at 168th whereas Vietnam ranked at 141st among 180 countries of the world in terms of environmental sustainability. Another point of similarity is that in both Vietnam and India spiritual/religious practices remain a strong formal institution (Vu & Tran, 2021). Therefore, it would be insightful to study how these spiritual practices are contributing to creation of novel and creative pathways of environmental sustainability in these contexts.

Lastly, because of weak legal enforcement in the developing contexts, informal institutions promoting ethical and moral orientations remain important (Vu & Tran, 2021), particularly in raising moral awareness toward environmental sustainability. Existing studies reveal that mindfulness practices are associated with a heightened commitment to ethical values (Gould, 1995), a sense of morality (Pandey et al., 2018; Shapiro et al., 2012), pro-social behavior (Donald et al., 2019; Fischer et al., 2017; Hafenbrack et al., 2020), and sensitivity to sociocultural norms (Chan & Ananthram, 2019).

Nevertheless, research on mindfulness in business ethics and organizational studies remains in its in infancy (e.g., Reb et al., 2020) with much room to be explored. Mindfulness adopted in organizations is often criticized as "instrumental" and devoid of an ethical focus (Qiu & Rooney, 2019). These mindful practices are often reduced to a stress reduction technique in order to serve organizational end purposes (Purser & Milillo, 2015; Vu et al., 2018). This approach is being critiqued by some scholars (Verhaeghen, 2015; Vu et al., 2018) who argue that when mindfulness practices are stripped off their ethical foundations, they may lead to undesirable behaviors (Monteiro et al., 2015). Some studies have also demonstrated how mindfulness programs are being increasingly employed by corporates as a tool to enhance individual productivity (Magid & Poirier, 2016). Such approaches however lead to a "heightened sense of self" and enhanced sense of self (Qiu & Rooney, 2019). This corporate mindfulness is counterproductive in nature and contradicts the objectives of traditional mindfulness practices, which aimed at eradicating a sense of "self." For instance, Buddhist mindfulness emphasizes an ego-less/non-self-state of existence whereby the self is decentered without products of ego (Epstein, 1988; Purser & Milillo, 2015) following the Four Noble Truths, identifying over-attachment to any phenomenon including the self as a source of suffering. Similarly, Sikh mindfulness training endeavors to achieve ego transcendence through a step-by-step process of mind training rooted in ethical discipline combined with meditative absorption and spiritual wisdom. It is a gradual process of realization and contemplation to convert the ego-conditioned mind (*manmukh)* to a nonconditioned self (*gurmukh*) (Panesar, 2017) through the practice of mindful contemplation (*Simran),* rightful thoughts (*sache vichar*) and rightful action (*shubhkarman*). These practices liberate the conditioned mind from the shackles of material desires (*maya*) and enable an individual to inculcate acceptance, compassion, and empathy toward the human and non-human world (Singh, 2005; Shackle et al., 2001). Such approaches highlight the limitations of secular mindfulness approaches, which are bereft of an ethical focus, embedded in tradition-specific mindfulness approaches.

We argue that secular mindfulness approaches have not been able to keep mindfulness values (e.g., integrity, ethical core, liberative perspective and intent, etc.) embedded in different traditions intact in an organizational context. In this paper, we respond to the need to further understand how mindfulness practices can promote moral orientations toward environmental sustainability, ethical choice (Kreplin et al., 2018), and other-focused ethical behaviors (Orazi et al., 2021). We explore the ethical dimensions of the mindfulness perspective-taking approach from the Buddhist and Sikh traditions in the context of Vietnam and India. We present a cross-cultural dialogue from an Eastern lens to show how mindfulness approaches in general can benefit from understanding diverse mindfulness practices embedded in different traditions that embrace other-focused ethical behaviors and promote sustainable practices.

8.2 Promoting Environmental Sustainability Through Morality

Concerns over climate change and rapidly declining biodiversity have highlighted the increasing importance of behavioral changes in developing effective mitigation strategies (McLoughlin et al., 2019; Poortinga et al., 2019). Moving toward behaviors promoting ecological sustainability is a moral issue (Nielsen & Hofmann, 2021). It is argued that morality can help an individual to distinguish between right and wrong behaviors, influencing a person's moral values in the process of process of moralization (Feinberg et al., 2019; Hofmann et al., 2018).

Increasing studies have linked morality and environmental behavior (e.g., Feinberg & Willer, 2013; Hurst & Stern, 2020; Jia et al., 2017; Wolsko, 2017). Morality is important for inculcating environmental consciousness since this involves sacrificing one's individual interest and incorporate concerns for environmental well-being (Nielsen & Hofmann, 2021). Some empirical studies have demonstrated that moral values and belief system translate into a sense of increased ecological responsibility among individuals (Bolderdijk et al., 2013). In the course of moralization (Feinberg et al., 2019; Rozin, 1999), value judgments are mediated through education, persuasion, and logical reasoning. Moral values are regarded as a motivational power for self-restraining desires (Mooijman et al., 2018). For instance, adoption of green products and avoidance of wasteful consumption involves the process or moralization reflecting on moral values (Broome, 2008; Jamieson, 2010; Markowitz & Shariff, 2012). Other studies have shown how individuals can be motivated to look at meat consumption as an "immoral" behavior by educating them about the adverse consequences of meat production and consumption for the natural environment and animal welfare (Feinberg et al., 2019; Poore & Nemecek, 2018).

8.3 Mindfulness and Morality

Awareness and cognition are important factors influencing ethical behavior (e.g., Jones, 1991; Lehnert et al., 2015; Rest, 1986; Trevino, 1986). Rest (1986) focused on four cognitive states of moral reasoning: awareness, judgment, intent, and behavior (Craft, 2013). Within Rest's (1986) model, awareness is the first step in understanding the moral reasoning process of ethical decision-making. The process starts with awareness of the ethical nature of choice in order to differentiate between right and wrong (Craft, 2013). Similarly, Lehnert et al. (2015) in reviewing ethical decision-making processes found that awareness of ethical issues is the first step toward an ethical act. Without being aware of the presence of an ethical issue, one cannot bring moral reasoning to bear on the issue while forming intentions and taking action. For instance, if one is unaware of the existence of an ethical issue, decisions are more likely to be made based on a cost-benefit analysis rather than consulting ethical values (Rest, 1986). Trevino's interactionist model (1986) also highlighted how an individual behaves when faced with an ethical dilemma. Such cognitions are influenced by traits (e.g., self-awareness) and the context (e.g., nature of the job, organizational culture). On the other hand, Jones (1991) emphasizes that awareness of the moral aspect of an issue, particularly the different levels of moral intensity, influences the imperative of a situation. In other words, antecedents of ethical behaviors include awareness and a heightened state of mindfulness (Ruedy & Schweitzer, 2010; Lehnert et al., 2015; Mihelič & Culiberg, 2019).

Mindfulness is situated in the functional domain of attention, cognition, emotion, and behavior (e.g., Good et al., 2016; Kang & Whittingham, 2010; Reb & Atkins, 2015) and is often characterized as a reflexive state of mind which enables individuals to engage in the present, to be aware of new things, and to be sensitive to context with distinction-making and differentiation (Langer & Moldoveanu, 2000). With such functions, mindfulness reflects awareness of morally relevant internal and external cues and fosters ethical behavior (Sevinc & Lazar, 2019) that can potentially enhance individuals' awareness of an ethical issue (Rest, 1986) and contextual cues influencing moral intensity (Jones, 1991). Studies on ethical decision-making claim that ethical errors often occur due to automaticity (e.g., Kern & Chugh, 2009); therefore, awareness of ethical issues is crucial to avoid mental pitfalls and framing effects (Kern & Chugh, 2009) that can influence moral implications of an act (Tenbrunsel & Massick, 2004; Lehnert et al., 2015). Studies show that mindfulness can provide a type of attention and awareness that can help to overcome limitations in judgment where individuals take into consideration not only the content but also the context of information (e.g., Robinson et al., 2017).

Studies on mindfulness have been increasing in the moral and ethical domain (e.g., Reb et al., 2020; Sevinc & Lazar, 2019; Sutcliffe et al., 2016), and most studies examine mindfulness as a trait (e.g., Gentina et al., 2020; Pandey et al., 2018; Reb et al., 2020; Small & Lew, 2021), referring to personal attitudes (Brown & Ryan, 2004, p. 837) that distinguished one person from another by the dispositional differences in mindfulness (Allen & Kiburz, 2012, p. 371). Scholars found that mindfulness is closely linked to ethics and moral judgment (e.g., Dhandra & Park, 2018; Kang & Whittingham, 2010), moral reasoning and ethical decision-making (e.g., Lampe & Engleman-Lampe, 2012; Pandey et al., 2018; Ruedy & Schweitzer, 2010; Shapiro et al., 2012), ethical behavior (e.g., Kalafatoğlu & Turgut, 2017), interpersonal justice perceptions (Reb et al., 2020), and ethical organizational culture (e.g., Kasser & Sheldon, 2009). Mindfulness also supports nonmaterialism values, well-being, and prosocial behavior (Fischer et al., 2017), reduces egocentric bias (Padey et al., 2018), and enables ethical decision-making through greater awareness of the environment, tolerance of situations, and honesty (Ruedy & Schweitzer, 2010).

Nevertheless, because studies on mindfulness when taking the secular approach in an organizational context often emphasize on moment-awareness, it departs from some meaningful tradition-specific dimensions of mindfulness. For instance, mindfulness – *sati* (smṛiti in Sanskrit) – in Buddhism transcends moment awareness but the word *sati* refers to the verb "to remember" (Anālayo, 2010; Gethin, 1998) and the importance of remembering the Dharma (Buddhist teachings) in practicing mindfulness (e.g., Purser & Milillo, 2015; Qiu & Rooney, 2019; Vu et al., 2018). These guide practitioners to attend to the ethical frameworks and compassion set out in the Dharma. Lacking that dimension in mindfulness training and studies has led to criticisms of secular mindfulness approaches in using mindfulness as a technique to serve organizational end purposes (e.g., using mindfulness as stress-reduction technique to facilitate employee productivity) (Purser & Milillo, 2015; Vu et al., 2018), questioning its ethical foundation (Verhaeghen, 2015; Vu et al., 2018). The compassion dimension of mindfulness that is embedded in tradition-specific approaches (e.g., Gilbert, 2019) has also been taken for granted since studies tend to link mindfulness to ethical decision-making with a focus on self-focused ethical behavior rather than other-focused ethical behaviors (Orazi et al., 2021). Such an approach tends to strengthen a sense of self and self-centeredness (Qiu & Rooney, 2019) rather than reducing it to eradicate egoistic desires leading to suffering, which is a foundation to mindfulness practice in traditions like Buddhism or Sikhism.

8.4 Buddhist "Right Mindfulness"

Right mindfulness (sammā sati) is part of the eight mutually reinforcing principles of the Noble Eightfold Path. It requires an individual to take into consideration the "right view" (e.g., understanding sources of suffering, the depending arising and impermanent nature of phenomena) to enable right effort and right action. A mindful "right view" involves the realization of "right action" and gives sense of an obligation to act on the right view (Purser & Milillo, 2015), similar to the notion of moral judgment and moral intent in moral reasoning (Rest, 1986; Small & Lew, 2021). Buddhist mindfulness is therefore embedded with right view and right action, reflecting the ethical dimensions in practicing mindfulness. The objective of right mindfulness in Buddhism is to eliminate suffering as demonstrated through the Noble Eightfold Path by enabling changes in behavior and psychological traits

(Purser & Milillo, 2015) and attaining wisdom (Krägeloh, 2016; Vu & Gill, 2018). Right mindfulness in Buddhism is the de-automatization of habitual reactions and perceptual evaluations (Purser & Milillo, 2015) that could potentially limit the cultivation of wisdom due to ignorance caused by over-attachment to habits and individual perceptions. To facilitate such objectives, right mindfulness is tied to an ethical and soteriological framework (Jinpa, 2019) that highlights the foundation of morality (Sīla) in refraining from causing harm in relation to others and the importance of developing wisdom and insights into the fundamental truths of a phenomena (e.g., impermanence, emptiness).

Right mindfulness can often appear as self-focused as the practice aims at attending to sensations and observing reactions of the mind to various stimuli and events that might seem to reflect self-improvement. However, it is through the accumulation of insights from the practice that individuals become less self-preoccupied, leading to the experience of selfless and non-self (Shapiro et al., 2012). It moves beyond the practice of focusing on self-improvement, the illusion of the self to self-liberation by not trying to make or force anything to happen but to let the process unfold naturally to discover thought without a thinker (Epstein, 1988) and seeing contents of the mind as impersonal events arising in awareness (Shapiro et al., 2012). This highlights how Buddhist mindfulness rejects the notion of self-centeredness in practicing mindfulness. Within the Mahayana tradition, compassion is a target for training in its own right (Gilbert, 2013, 2019). Furthermore, the theory of emptiness (Sunyavada) in Buddhism highlights that all earthly phenomena including our own self are empty in intrinsic terms (Thich, 1999; (Sahdra et al., 2010).

Right mindfulness does not just emphasize on moment awareness but engages with a wider range of perceptual, cognitive, and affective phenomena (Bodhi, 1994; Kudesia & Nyima, 2015) through an emergent experience (Kudesia & Nyima, 2015). Therefore, Buddhist right mindfulness is (i) situated within the Buddhist ethics (guided by right view and right action in the Noble Eightfold Path) and based on the foundation of morality (Sīla); (ii) part of the process of attaining wisdom and insights on the nature of the self and phenomena within the journey toward enlightenment; and (iii) embedded in compassion with other focus rather than self-focus orientations.

8.5 Sikh Mindfulness

Mindfulness practices are incorporated in the Sikh holy book, Guru Granth Sahib, as a soteriological practice aimed to overcome a "state of spiritual imbalance" (Kalra et al., 2013) through a deep transformation of the mind, rooted in psychology, ethics, and the idea of community. Mindfulness is a diagnostic tool that "promotes moral awareness, empathy and humility through cognitive development of psychological characteristics and ethical conduct" (Kalra et al., 2013: 196).

> The mortal's business is with his own mind. The one who disciplines his mind attains perfection (SGGS: 341).

The penultimate objective of mindfulness practices within Sikhism is to conquer ego (hāumāi) through "loss of self" and alleviate suffering (dūkh) rooted in egocentric desires (Mandair, 2012). Sikhism propounds that continued focus on the self gives rise to a host of destructive, psychological patterns where in the mind "plots, calculates, desires, manipulates, flares up in anger and indulges in waves of negative emotions" (Mandair, 2012). Sikh mindfulness provides guidance to acquire wisdom by acting on a righteous manner and eliminate suffering embedded in greed, covetousness, lust, anger, and animosity (Singh, 2005). Sikh mindfulness is not just "one of the many activities in life, but an orientation at the core level, involving the total self" (SGGS: 128). Unlike Western mindful practices, the aim of Sikh mindfulness is not merely a stress reduction technique (Purser & Milillo, 2015) but a method of transformation of the human mind-shifting focus from one's extrinsic self to the "divine light" within through a constant process reflection and self-realization in order to achieve a "symbiosis between the material and the spiritual aspects of life" (SGGS: 128). The Sikh holy book, Guru Granth Sahib, lists a step-by-step process which enables an individual to acquire wisdom and live in an enlightened way. This mindfulness perspective is centered around the practice of continuous contemplation (naam Simran) and selfless service (seva). These two practices are deeply intertwined and regarded as the pathway to achieve a state of mental equilibrium through transformation of a manmukh (an ego-conditioned state) to a gurmukh (virtuous non-ego-conditioned state) (Singh, 2005).

One of the fundamental practices of Sikh mindfulness is naam Simran (continuous contemplation). The word simran is equated to multiple interpretations ranging from "remembrance" (Kalra et al., 2013) to "reflection" (Sahota, 2011), "spiritual meditation" (Nesbitt, 2005), and the "art of stillness" (Singh, 2004). Guru Nanak, the founder of the Sikh faith, emphasized on Simran as a holistic approach of "going within the self" to achieve a sense of equilibrium. He said, "Meditating, meditating in remembrance I have found peace" (SGGS: 202). The concept of Simran goes beyond moment awareness, and "non-judgment attention to the present moment" (Purser & Milillo, 2015). Instead, it involves a process of deep contemplation of one's past experiences and memories to reflect on one's wrongdoings and weaknesses in order to achieve self-realization and acknowledge one's dereliction (SGGS: 651). The practice of Simran provides one with the necessary support mechanism to eradicate oneself from the chains of negative thoughts and impulses and replace them with righteous thoughts (sache vichar) to achieve a "perpetual state of mental equilibrium," through the integration of concentration and insight (vichar) (Mandair, 2012). It is rooted in an ethical compass that discerns right from wrong (Scott, 1997). Sikhism emphasizes how the practice of Simran is enhanced in a congregation (sãngat) of like-minded individual. As each participant meditates on past and present experiences, "there is thread that weaves through each person, and the uniformed thread increases the magnitude of the power of Simran" (Kalra et al., 2013:197).

> Seek the sanctuary of the saints and fall at their feet; your suffering and darkness shall be removed (SGGS:51).

While Simran is the key to acquire self-awareness, Sikh ethos also lay strong emphasis on righteous actions (sach achar) toward one's fellow being and community service (seva) as the necessary means to achieve mental peace and equilibrium (Panesar, 2017). Sikhism advocates that the practice of seva enables individuals to cultivate humility and transcend the ego to "see beyond themselves" (Panesar, 2017). It promotes realization that we are not divorced from others around us (SGGS: 128). Social responsibility through cultivation of "righteousness" (sach acha), "good deeds" (shubhkaram), and "compassion" (daya) remain core concepts in Sikhism (Singh, 2020).

Thus, from the above discussion, one can conclude that Sikh mindfulness aims at a transformation of one's psychological framework through a step-by-step process of mind training based on (i) constant contemplation and reflection (Simran) to achieve self-awareness (ii) embedded with Sikh ethics of right thoughts, right action, and right conduct (iii) intrinsically connected to service to humanity rather than a self-focused orientation.

8.6 Context of the Study

In order to advance a cross-country Eastern dialogue on mindfulness practices toward environmental sustainability, we recruited participants from two different traditions across two geographical settings: the Mahayana Buddhist practitioners in Vietnam and Sikh mindfulness practitioners in India. The Mahayana tradition within the engaged Buddhist movement in Vietnam (Vu & Tran, 2021) is rooted on the application of Buddhist principles and values to resolving contemporary socio-economic and environmental challenges (Thich, 1999). Examining practitioners from the Mahayana tradition also helps to illuminate other-oriented approaches in mindfulness. The Mahayana practice embodies an ethical system, rooted in the principle of "seeing things as they are" (Vokey, 2001). Similarly, Sikh mindfulness practices are strongly rooted in Sikh ethical principles and cultivation of spiritual wisdom while freeing oneself from the materialism, avarice, and worldly desires, rooted in ego-centrism (Singh, 2005). Sikh mindfulness does not only focus on self-awareness through mindful contemplation but is strongly embedded in an "other-focused" perspective drawing from the ethical principles of righteous actions toward fellow human beings and giving back to the community (Singh, 2005). Analysis of Sikh mindful practices thus enabled us to understand how an other-focused, community focus may be incorporated in mindful practices.

8.7 Methodology

We adopted a qualitative research design (Creswell, 2013), rooted in an interpretative framework. Firstly, we wanted to advance a comprehensive and nuanced analysis of other-focused ethical traditions in Sikh and Buddhist mindful practices and avoid the codification and simplification dominant in quantitative methodologies (Felver & Jennings, 2017; Pace, 2013). Secondly, it enabled us to understand the multiple mechanisms through which our participants mediated their mindful practices without reducing them to a "one size fits all" framework (Bluth et al., 2016; Dariotis et al., 2016). Third, we were able to analyze how the participants' ontological (nature of reality) and epistemological beliefs (how we know what we know) (Lincoln & Guba, 1985) were conditioned by their religious/spiritual doctrine, which in turn influenced their attitudes toward sustainability.

We adopted an exploratory research design from a phenomenological-hermeneutic perspective (Cohen et al., 2018). Phenomenology is defined as the study of human existence as it is implicitly lived and experienced by participants (Moustakas (1994:74). This approach incorporates the perspective and lived experiences of the research participants in "all its richness" in contrast to reporting only objective findings (Smith et al., 2009). Phenomenological method helps the research to conduct an in-depth evaluation of individual cases by giving "rich details" and "thick descriptions" (Gill, 2014; Cope, 2011). An added advantage of a hermeneutic approach is that it helps to recognize that the participants' perspective is one of the many interpretations of a given phenomenon (Felver & Jennings, 2017). Accordingly, in this research design, a phenomenological-hermeneutic perspective was used to make sense of how participants channeled an other-focused orientation in their mindful practices.

We adopted theoretical sampling in this research, wherein the selection of participants was based on whether they could provide insights into the phenomena under scrutiny (Corbin & Strauss, 2008). In the Indian context, we recruited participants who had had been practicing Sikh mindfulness for at least the past 2 years and were engaged in pro-environmental initiatives at the ground level. In Vietnam, participants were gathered through snowball and purposeful sampling. The aim was to recruit participants who have been practicing and applying mindfulness practices in the work context for more than 5 years.

8.8 Buddhist Right Mindfulness

Buddhist practitioners in our study highlighted a number of different approaches they applied that facilitated other-focused orientations in mindfulness practices that influenced their awareness of environmental sustainability: decentering mindfulness and compassion mindfulness. Below we demonstrate the different forms of Buddhist right mindfulness practices, the tensions associated in the practices, and implications of the practices for moral reasoning toward sustainability.

8.8.1 Decentering Mindfulness Approaches

A number of Buddhist participants highlighted decentering as an important mind-training aspect in the way they practice right mindfulness. Decentering as our participants explained refers to the ability of practitioners to move away from a within one's subjective experience onto experiencing phenomena as they are:

> I use mindfulness not just for meditation but to question my 'self' and my 'perspective' in how I see and understand things. For me, as a manager, it is important to understand people and situations to reflect on my decisions. That is why for me, mindfulness is a decentering (*phá chấp*) practice that enables me to see things as they are rather how I want them to be, how I believe they should be or how I expect them to be [...] it helps a lot question my managerial decisions, particularly in situations when deciding what is right and wrong is conditioned by the characteristics of who is involved and the context [...] Decentering mindfulness for me involves both meditation (mind-body dialogue) and contextual reflection based on internal-questioning dialogue [...]. (BR6)

Decentering mindfulness is described as a dynamic approach that includes both meditation techniques and internal dialogue and reflection based on Buddhist philosophical teachings to avoid over-attachment to subjective interpretations that may block understanding of phenomena and influence decision-making. It reflects disidentification from internal experience and a process of stepping out from one's experience (Bernstein et al., 2015). This description aligns well with theoretical descriptions of decentered mindfulness (Lau et al., 2006) and the Mahayana meta-ethics that places important on understanding the conditioned nature of propositional truth with an intrinsic goodness that acknowledges others rather than the "self" (Vokey, 2001). Decentering mindfulness in our findings included two main approaches: detached mindfulness and self-distanced mindfulness.

8.8.1.1 Detached Mindfulness

Our participants demonstrated detached mindfulness (*phá bám chấp*) as a practice that enables them to step away from conceptual analysis in the absence of a goal-oriented outcome and subjective perception/emotions to allow objective interpretations of thoughts and phenomena:

> Mindfulness for me is a way to decenter from my own thoughts and feelings to be able to see issues and problems in their own nature... First, while I do routine things without the need to put effort in it, I try to reflect on and identify the issue that I'm facing such as considering to add an item in our project to satisfy local authorities. Second, I try not to involve my own thoughts of how I feel about that request but consider the issue on its own context, the personnel involved, our capability in fulfilling the item and impacts on the overall project. Lastly, I revisit the feasibility and the overall potential outcome of the project with an objective view rather than comparing it to the intended or expected outcomes. Because we are initiating the project to help local communities build recycling facilities, the final outcome should consider the overall impact to enhance environmental sustainability for local communities. If I would not have gone through this process, I would have got involved

> personally and I would not feel comfortable to continue the project by considering favours or paying extra for local authorities. (BR16)

The step-by-step explanation of the process of detached mindfulness is somewhat similar to Well's (2005) metacognitive model of detached mindfulness and Rest's (1986) process of moral reasoning. Our participant highlighted the following: (1) meta-awareness (awareness of moral issue in doing favors for local authorities); (2) cognitive decentering (detached from personal thoughts/emotions on the moral issue influencing judgment); and (3) low goal-directed coping (detached from the need to avoid/reject/control or suppress the threat of the moral issue to attain a specific goal, instead attending to the issue context-sensitively with flexibility). The role of community and sustainability orientations was emphasized as the key compass within the process, which highlights the other-focused practices in the Mahayana tradition. This was further demonstrated by a participant who used walking meditation to facilitate internal dialogue for detached mindfulness:

> I usually do walking-meditation if I have the opportunity to do so. While I walk, I initiate my thoughts of the problem that is bothering me. I had to decide on withdrawing fund and collaboration from a project with a strategic partner of the company because of our financial situation…it can affect the quality of imported materials to commit ourselves to be environmentally sustainable with our waste…I felt guilty, anxiety and were hesitant to make the final decision. […] During walking meditation, I raised a series of thoughts on situation, trying to hold on to them simultaneously when I focus observing the things happening in the street around me as I walk by. That helps me to realize I don't live by thoughts as they are impermanent and are passing by. (BR8)

Recognizing the impermanent state of thoughts helps individuals to unlock attachment to negative thoughts and events. It enhances objective awareness of an event/problem where attention remains fluid or perceived conceptualization of meanings to facilitate moral reasoning without negative thoughts. Some participants illustrated different meditation techniques to attain detached mindfulness in responding to moral issues on sustainability. For instance:

> I use imaginary meditation for detached mindfulness in dealing with difficult situations to make sure that my own perceptions are not influencing the real state of the issue. […] For instance, I was aware of a misconduct of a colleague…who changed our distributor ignoring our commitment to environmentally friendly input materials… I used imaginary meditation to look for an answer. I imagined my thoughts as clouds so it will be useless for me to push the clouds away or try to grasp them. I allow thoughts like fear of being excluded, fear of being neglected, fear of being a coward by not speaking out, fear of guilt and so on. Once they passed by, I realized whatever others think or may think is something beyond my control. I let go of my fear and reported the case. […] it turned out the case was handled confidentially and it did not affect our work […]. (BR22)

The above participant used imaginary technique to detach from imagination of scenarios that caused fear in dealing with moral issues. Detaching from fear here enables individuals to avoid maladaptive elements involved in cognitive and emotional (Kross et al., 2005) processing that could have clouded moral judgment and behavior toward sustainability issues.

8.8.1.2 Self-Distanced Mindfulness

A number of respondents highlighted the effectiveness of self-distanced mindfulness as an approach to facilitate reflexivity, criticality, and objectivity in moral reasoning for environmental related issues by eliminating self-focused thoughts:

> I use self-distanced mindfulness *(phá chấp ngã)* to facilitate my practice and understanding of non-self in dealing with challenging issues. For example, I engaged myself in a self-reflexive dialogue to reveal if my perspective and thoughts were influencing our initiatives to invest in a large sustainability project. My experience from previous projects was not so good as we had a really difficult time with local officials. There were a series of requests to approve the project, including paying transactional cost. Still, the outcome of the project was not as expected [...] In self-distanced mindfulness practice, I often ask myself three types of questions: (1) what was my intent when I made the decision? (2) how did I feel about the outcome of the decision? and (3) why did I feel that way? If any of my answers involved the role of 'self' or ego in it, I knew that I was biased by my own perspective and thoughts and should reject them in considering this new project. [...] I felt bad about the previous project because the outcome was not up to my expectations and commitment to partners...our environmental index of the project was not fulfilled...that affected my reputation and I had to involved in lobbying activities that I did not feel comfortable about, not because of the overall outcome of the project. (BR35)

The self-distance mindfulness approach that respondents highlighted in our study is not about rejecting past experiences to embrace moment-awareness but to "remember" Buddhist teachings on non-self and learn from past experience to enable an ego-decentered perspective (e.g., Libby & Eibach, 2002; Robinson & Swanson, 1993). This approach facilitates a move away from self-immersed to self-distanced perspective with wide attentional breadth (Dane & Brummel, 2014) to phenomena occurring both internally and externally (Dane, 2011) following Buddhist principles and metaethics in practicing mindfulness (Anālayo, 2010; Gethin, 1998). Participants also used self-distanced mindfulness to engage in the of process of asking-why to understand organizational intent of a moral issue with a more objective awareness:

> After we close a deal, I always reflect on it to take away some lessons if I can. I do it with mindfulness to distance myself from being a member of the company and ask a series of 'why' questions to gain insight and closure of a deal [...] Why did we engage in that deal? Why were there complaints from stakeholders? Why was the outcome not as expected? [...] this helped me to learn that sometimes the intent of our company affected the outcome of the deal, not because we had difficult stakeholders or because of the context as the company claimed. We promised to handover a fully equipped and functional factory...with fully equipped recycling facilities but because we were engaged in multiple construction projects simultaneously with the intent to be efficient, the quality we delivered was not as promised. [...] Such reflection for me was useful to take away and raise quality and moral issues in implementations in our next projects and as a Quality Assurance manager [...]. (BR28)

Asking-why was a strategy that some participants used in self-distanced mindfulness to gain objective insights into a phenomenon and organizational intent without promoting avoidance in acknowledging the negative implications for future lessons. Adopting the why analytical perspective, the participant was able to take a step back and look at his past experience through a distant lens (Kross & Ayduk, 2011). This approach is similar to the distanced-why strategy used to enable individuals to construe experiences in abstract terms to meditate emotional reactivity (Kross et al., 2005). However, in our case, distanced-why strategy in self-distanced mindfulness facilitated understanding of organizational phenomenon without individual bias as being part of the company by reflecting on their past, modifying their initial conceptualization on the basis of new contextual information, and submersing negative responses in the future (Foa & Kozak, 1986).

8.8.2 Compassion Mindfulness Training

In sharing their experiences of using mindfulness to tackle moral issues of environmental sustainability, participants stressed the importance of compassion mindfulness training as part of their Bodhicitta (awakening mind striving toward cultivating compassion for the benefit of all sentient beings) journey. Participants emphasized on the other-focused rather than self-focused benefits in practicing mindfulness:

> Compassion mindfulness training is very useful for me to remind me of reviewing the intent in my actions. I engage myself in an internal-dialogue that helps me to compare motives in different actions. For instance, do I oversee to report a mistake out of compassion or to gain reciprocal relationship? Do I expect anything in return?...Does it affect the whole project and the outcome for the community? For me, engaging in such dialogues over time trains my mind to be aware of and judge any motive leading to a decision or behaviour based on compassion. [...] it has proved to be helpful for me in a sense that my actions are activated and judged based on motivation to help others and the overall outcome of our projects, not my 'self' [...] that is part of the experience of non-self. (BR26)

Participants highlighted how compassion mindfulness training can facilitate an understanding of non-self and moral reasoning (Rest, 1986) in promoting moral orientations for sustainability projects. This approach reflects the Mahayana practitioners in practice by emphasizing good deeds and dedicating karmic merit to others, not the "self," which they consider as part of their mindfulness practice. For instance:

> I practice it through insight meditation (*Vipassanā*) (*thiền minh sát*), by settling my mind to gain insight into the true nature of my mind without the 'self' to enable compassion to arise naturally... If I can accept uncomfortable acts that I'm not proud of like bribery but with the intention to deliver social benefit to the community, I am dedicating my good deeds to others without caring for my on karmic consequence. However, if I use the name of compassion to justify unethical acts all time with the intention to get things done that is no longer compassion. Gaining insights to my mind helps with that process. (BR19)

The above participants highlighted how compassion mindfulness training does not only involve compassion but the need to consider the foundation of morality in the practice to gain wisdom through systemic thinking to avoid prestige seeking or self-interest actions that can mistakenly be justified by the practice of compassion without wisdom. This finding highlights the central importance of the need to "remember" fundamental principles/teachings rooted in mindfulness practices (e.g., Kabat-Zinn, 2019; Purser & Milillo, 2015) to regulate competing motives (Gilbert, 2019) in moral actions with wisdom to avoid the instrumentalization of mindfulness practices in moral reasoning.

8.9 Sikh Mindfulness Practices

Sikh participants in our study elaborated on different dimensions of their mindful practice that facilitated them to incorporate heighted environmental awareness in their moral reasoning and decision-making processes: (a) self-distanced mindfulness, (b) empathy-based mindful practices, and (c) community-oriented mindfulness. We elucidate on these dimensions in detail in the subsequent sections, along with the struggles in implementing these practices, and the resulting implications for environmental awareness/consciousness.

8.10 Self-Distanced Mindful Practices

A number of respondents in our study highlighted how mindfulness practice gave them the necessary pathway to move beyond their own thoughts, concerns, desires, and cognitive biases and rid themselves of ego-centric desires. According to Bernstein et al. (2015), self-distancing involves distilling of your thoughts and past experiences from your own judgment and the stories you weave around these internal experiences. This is reflected in the following quote from an interview participant, wherein he elucidates on how the daily practice of *naam Simran* helped him to develop an objective view of the long-term sustainability of his project using self-distanced mindfulness:

> Mindfulness for me is a means to distil my thoughts and emotions and look at issues from a fresh perspective, without being blinded by extreme emotions like anger, resentment or jealousy. I feel that nothing blinds us more than when we approach an issue with anger in the heart. Over the years, I have developed a daily ritual, where in I get up at 4 am which is regarded as the ambriosal hour in Sikhism and first do chanting of the *naam* while emptying my mind of all thoughts. Over the years, this practice has helped me to treat my thoughts and feelings in an objective way. While reflecting on the happenings at work, I am able to keep my own thoughts, and feelings about the persons' and things involved at bay. After doing this practice, I become clearer about my strategy and action plans for the future. Mindfulness has helped me to be sincere with people around me, about what I can do, what is my specialty and what I would not be able to achieve in the given time frame and budget. It also helps me to be grounded, and not be pompous about what I can do or not do. I used to have an ambition that I wanted to make a name for myself, but that is no longer my concern. Doing good for the society and environment is more important. (SR1)

This interview quote demonstrates that by stepping back from his thoughts and feelings, the respondent displayed greater self-awareness of his own capacity, as well as heightened moral awareness (Rest, 1986) which then translated into heightened ethical consciousness of sustainability issues in his moral reasoning. We also see how self-distanced mindfulness provided the interviewee with a mechanism to rid themselves of preconditioned tendencies (Garrigan et al., 2018) such as "being pompous." This then translated into enhanced moral decision-making, wherein the interviewee elucidates on the importance of "doing good for the environment and society." This resonates with the "other focused" orientations in Sikh mindful practices, elaborated upon in the previous section.

The Sikh concept of mindfulness is not limited a particular posture or time of the day but is conceptualized as a continuous practice of contemplation that should be conducted through the day. Our interview participants revealed that they integrated different practices, according to what they found appropriate in different situations. This is reflected further in the following quotes from participant who practice mindfulness through *walking simran* meditation:

> I have been practicing Walking Simran Meditation since the past nine months. With each step I take, I do *Simran* of God's name, and listen to *kirtan* (devotional music). Since I began this practice, I have begun to feel a closer connection to the One. I have begun to keep *Waheguru* (God) in my remembrance at all times. It has changed my way of thinking. During my mindfulness practice I go back to my past experiences and see whether my intentions had been motivated by ego. I was able to self-reflect on how such desires led me to act in ways where I only considered short-term gains and my own greed, without any concern for others. While I regret some of these decisions, I take it as a learning process going forward. Mindfulness for me is like an alarm which goes off in my head, whenever I deviate from the right path and indulge in greedy behaviour or tell lies. I am not perfect. I still make mistake. But I am trying my best. (SR4)

The above quote illustrates how a self-distanced perspective enabled the respondent to achieve a heightened moral awareness of negative impulses in the past, rooted in egoism, greed, and lack of concern for others. By using a self-distanced mindfulness approach, the respondent was able to strip these experiences of unnecessary tangible details and take the bigger picture into account in their reminiscences (Kross & Ayduk, 2011; Petrova et al., 2021). This phenomenon has been christened as "reperceiving" wherein an individual stands back and observes his/her thoughts and experiences rather than submerging oneself in those experiences (Shapiro et al., 2012). Mindfulness, rooted in Sikh ethos, provided the respondent with a reflexive mechanism for moral reasoning (Garrigan et al., 2018) and helped them to locate the basis of these impulses in their ego-centric desires. We also see that the mindfulness practice, as demonstrated by the interviewee, did not focus on "moment awareness" and rejection of past experiences. Instead, the interviewee went through a process of deep introspection and reflection to achieve enhanced self-awareness, rooted in the Sikh practice of *Simran.* This practice enabled the participant to extend beyond self-focused needs to display a profound concern for others, as well as long-term sustainability in their moral reasoning, instead of focus on immediate gratification of their needs (Singer & Klimecki, 2014).

8.11 Empathy and Compassion-Based Mindfulness Practices

While elaborating on their experiences of using mindfulness to deal with moral dilemmas, the participants stressed on the importance of empathy and compassion as a part of their mindfulness practice. A number of participants stressed on how mindfulness provided them with a scaffolding approach to look at issues from another's perspective and incorporate flexibility and self-reflexivity in their moral reasoning, as well as a heightened sense of compassion for the natural environment:

> Developing compassion and empathy has been the most useful aspect of my mindless practice. For me, mindfulness has essentially involved cultivating *chetna* (insight) and rooting my actions in righteous thoughts. I have been doing meditation for many years now. But recently I added a new practice in my daily routine. I spend 10–15 minutes every morning, reading a daily *hukam* (daily affirmation) from the Guru Granth Sahib and reflect on its meaning. This has given me a lot of clarity and enabled me to look beyond my own problems and frustrations. I have realised that how interconnected we are with the world around us. In the past I did not look at environmental regulations beyond what was compulsory under law. However, I have now developed a heightened awareness of how we are compromising nature while pursuing profits at any cost. I realised what use is profits and wealth if we leave nothing behind for the benefit of our future generations. (SR2)

The above quotes reveal how empathy and compassion in their mindful practice enabled these individuals to transcend their own thoughts and exhibiting a deeper concern for others, including the natural environment in their moral reasoning (Rest, 1986). Emotional awareness through mindfulness emerges as the cognitive mechanism through which these individuals were able to achieve a heightened sense of empathy and compassion in their moral reasoning. This confirms findings from existing literature which states that mindfulness helps to achieve enhanced emotional regulation and self-control which forms the underlying basis of other focused ethical behaviors (Sosik et al., 2019; Baer et al., 2019; Shapiro et al., 2012). The interview quote also reveals how compassion and empathy-based mindfulness practices can have other benefits such as cultivating a heightened awareness of the environment-economy trade-off. The above quotes also highlight how compassion and ethics-based mindfulness training was essentially rooted in wisdom and insights provided by other-focused Sikh ethical principles such as "cultivation of "righteous conduct" toward one's fellow beings (Singh, 2005). These point out to the critical importance of ethical principles in mindfulness practice (Purser & Milillo, 2015) in cultivating heightened ecological consciousness.

8.12 Community-Oriented Mindfulness Practices

Our participants revealed how they had incorporated a strong community-oriented motivation in their mindful practice, drawing from Sikh ethical principles. For instance, participant 5 reflected on how his mindful practice made him more conscious about local environmental issues:

> I believe that practising *nitnem* (daily prayers) and studying about Sikhism endowed with the consciousness to care for the environment. In Sikhism, nature is equated to the creator. Caring for the environment has become a form of worship for me. I try to avoid harm to the environment as much as possible. In my school we teach students the concepts of interconnectedness through the principle of pavanu guru̥ pā nı̣ pitāmātādhā rati mahatu (air is our teacher, water our father and the great earth our mother) which is enshrined in the *Guru Granth Sahib.* We also initiated a number of projects at the community level, such as celebration of Sikh environmental day and tree plantation. (SR 5)

The above participant reveals how his mindfulness practice was conditioned by Sikh principles of interconnectedness between the human and non-human world (Singh, 2020), which gave him the rationale and internal guidance toward incorporating concern for environmental issues in his moral reasoning. This is unlike secular mindfulness studies which argue that pro-social behavior is mediated by perspective taking and positive emotions such as compassion and empathy (Orazi et al., 2021; Grossman, 2015).

Another participant highlighted how community oriented mindful practice through the Sikh concept of *sangat* (congregation) provided her the spiritual guidance to bring in a community-oriented focus and a deeper concern for others in her moral reasoning:

> I started doing *Simran* in an *e-sangat* group last year during the lockdown. This practice helped me a lot during the lockdown. As Sikhs we are not only mandated to pray to better ourselves, but uplift the community with our ethical conduct. *Seva* and *Simran* are two halves of the path to spiritual upliftment. It is important to help people in the community who do not have what we have without expecting anything in return. What good is money if it cannot be used for the benefits of the underprivileged sections of society? As Sikhs, sharing our wealth and giving is a part of righteous living. Our *sangat* group began a *langar* (food preparation and food distribution) in the early months of the lockdown for construction workers who had lost their jobs and were struggling. In this process I achieved an internal sense of peace, and learnt not to be preoccupied with my own problems and thoughts all the time. (SR 3)

This participant elucidates on how a community-oriented enabled mindfulness training enabled him to transcend self-boundaries (Reed, 2009) and led to an intensified moral awareness of "others" and the "community" in their reasoning. Community-oriented mindfulness altered the individuals' life perspective, so that they were able to incorporate an other-focused perspective such as "helping the underprivileged" and "caring for the environment" in their moral judgment, leading to subsequent moral action. This approach was conditioned by the Sikh practice of *seva* (service) and working for community welfare, not just self-improvement (Singh, 2004), which the participants regard as an essential part of their mindful practice.

8.13 Discussion

Through our case studies, we moved beyond the focus on well-known secular moment-awareness mindfulness practices to highlight how mindfulness practices in specific-traditions (e.g., Buddhism and Sikhism) reflect dynamic approaches grounded in ethical foundations to navigate moral dilemmas in organization to enhance environmental sustainability projects. Our chapter moves beyond secular orientations in mindfulness practices to show how other-focused ethical behaviors and decision-making (e.g., Kreplin et al., 2018; Orazi et al., 2021) were facilitated by dynamic Buddhist and Sikh mindfulness practices (Fig. 8.1).

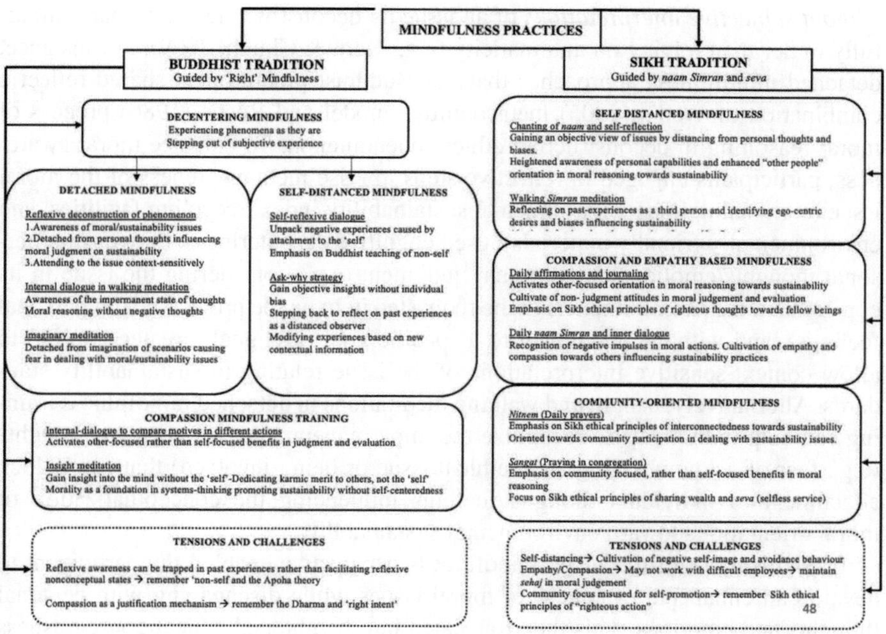

Fig. 8.1 Ethical dimensions of Buddhist and Sikh mindfulness practices toward sustainability. (Source: own compilation)

The mindfulness approaches we introduce in this paper reflect a perspective-taking approach grounded in moral, philosophical, and theological principles of these traditions. This approach extends studies in business ethics that often position and examine mindfulness as a trait (e.g., Gentina et al., 2020; Pandey et al., 2018; Reb et al., 2020; Small & Lew, 2021). For instance, the Buddhist decentering mindfulness practices aim at guiding practitioners to step out of subjective interpretations to experience phenomena as they are. To facilitate such approaches, participants in our study relied on the notion of non-self and the Mahayana metaethics that places intrinsic goodness and other-focused orientations at the center (Vokey, 2001) to guide them in deconstruct moral issues from the perspective of a distanced observer (Kross & Ayduk, 2011). Likewise, our data illustrates how Sikh mindfulness practices provided participants with a cognitive mechanism to reflect on their past-experiences and thoughts as a distant observer (Bernstein et al., 2015) and locate the basis of egocentric desires and negative impulses in their past experiences and actions. The participants' subjective experiences demonstrate that their mindfulness practice was conditioned by Sikh ethical principles with emphasis on honesty, rightful conduct, and an intrinsic balance between material ambitions and spiritual upliftment.

We introduce dynamic mindfulness practices from both these traditions as mechanisms to navigate moral orientations toward ethical decision-making promoting environmental sustainability. First, mindfulness *can enhance moral awareness without subjective interpretations* of an issue by deconstructing the scenario mindfully rather than relying on automaticity (e.g., Kern & Chugh, 2009). For instance, detached mindfulness approaches that our Buddhist practitioners shared reflect a combination of Well's (2005) metacognitive model and Rest's (1986) process of moral reasoning in deconstructing ethical phenomenon. To enhance moral awareness, participants engaged in reflective steps to raise meta-awareness of the moral issue (e.g., maintaining environmental sustainability index, recycling facilities, and environmentally friendly materials), used cognitive decentering to derive from personal thoughts/emotions clouding any judgment (e.g., considering the issue in its con context, the overall impacts of the requested item on the project, not on personal feelings), and without attachment to a specific goal (low goal-directed coping) to allow context-sensitive interpretations of the issue relating to sustainability standards. Alternatively, insight and walking meditations in detached mindfulness training can help individuals to recognize the impermanent states of negative thoughts (e.g., fear of acknowledging an unethical issue or being involved) that can unlock attachment to individual thoughts/feelings influencing the conceptualization of moral orientations toward environmental sustainability.

In Sikhism, a self-distanced mindfulness perspective enabled the participant to develop an enhanced awareness of moral issues while disengaging with personal thoughts and emotions. This then translated into heightened ethical consciousness in their moral judgment (e.g., being honest) and subsequent moral actions (e.g., working for the community). We also saw that the community-oriented focus in Sikh mindfulness practice provided participants with the spiritual guidance to transcend their own concerns and develop a deeper "other focused" perspective in their moral reasoning (e.g., concern for underprivileged, avoiding harm to the environment) leading to pro-social behaviors (e.g., local cleanliness drives, organizing food rations for the poor during COVID-19 lockdown). It is important to highlight that in both traditions, moral awareness was guided by other orientations of the traditions (e.g., the Mahayana metaethics in Buddhism and the Sikh ethical principles such as righteousness toward fellow humans, interconnectedness among nature and humans, and service to humanity) to emphasize on the importance of community considerations in considering environmental sustainability.

Furthermore, we found that when individuals practice compassion mindfulness practices in the scope of our study within the specific traditions of Buddhism and Sikhism, they do not engage in a cost-benefit analysis but relied more on ethical values (Rest, 1986) with other-focused orientations guided by the traditions rather than self-centeredness in navigating individuals' moral judgment and reasoning toward environmental sustainability issues. Mindfulness practices in our study contribute an alternative way to navigate moral intensity (Jones, 1991) in moral reasoning toward environmental sustainability. With other-focused orientations in the mindfulness practices that we introduced in both traditions, mindfulness can navigate moral intensity in actions in terms of *magnitude of consequences* of pursuing environmental sustainability projects since actions and motives are considered on the basis of community orientation and social well-being. For instance, Buddhist mindfulness practices were guided by the Mahayana metaethics that focuses on dedicating karmic merit to others (Clayton, 2009) through sustainable environmental projects. In Sikhism, the participants were guided by the principles of selfless service to humanity through *seva* (Mandair, 2012) and the principles of interconnectedness between nature and humans (Singh, 2020). This marks a departure from studies that often refer compassion mindfulness to self-compassion (e.g., Neff & Dahm, 2015; Van Dam et al., 2011) focusing on self-kindness, taking a stance of a compassionate "other" toward the "self," and recognizing negative thoughts toward the "self-concept." Our study highlights the importance of other-compassion mindfulness practices influencing moral reasoning promoting sustainability approaches.

To introduce tradition-specific mindfulness interventions in organizations to promote sustainability, we highlight some important boundary conditions. In Buddhist mindfulness practices, for instance, in decentering mindfulness training, the aim is to enhance reflexivity and reject subjectivity in observing organizational phenomena. However, often, in trying to solve problems and seek answers, practitioners tend to refer back to past experiences to categorize phenomena into familiar categories to identify what to avoid. That can lead practitioners toward dualistic or false conceptual awareness of phenomena that is based on reductionism rather than a nondual view to experience phenomena as they are. In the context of business ethics, reductionism or exclusion to try to fit a phenomenon in a familiar category is problematic since what is right and wrong have contextual extensions known as moral relativism (Brogaard, 2008), and moral decisions cannot be made without contextualization (Tsoukas & Cummings, 1997). Buddhist participants emphasized on remembering the notion of non-self while practicing mindfulness to remind practitioners how conceptualizations and phenomena are empty of intrinsic existence. Participants also highlighted how referring back to the Apoha theory[1] helps them to recognize how concepts and categorization involve exclusion and there is no exclusion that is ultimately true (Dunne, 2015). Therefore, the dissemination of tradition-specific mindfulness practices (e.g., decentering mindfulness training) involves an understanding of philosophical underpinnings of the tradition guiding mindfulness practices that might be time-consuming to introduce or might be counterproductive in some contexts where it might clash with philosophical teachings of individuals from different cultures or belief systems. On the other hand, without the dimension of wisdom in mindfulness, any realization of non-self and momentary experience of the impermanent and dependent arising nature of phenomena have almost no chance to develop (Kabat-Zinn, 2019). Without wisdom, the ability to discern judgment and appropriateness of action (Bierly et al., 2000) and acknowledge ignorance and limitations in knowledge (Nonaka et al., 2014), compassion mindfulness training might not work and can even become an instrument and justification mechanisms for immoral behaviors.

[1] The theory facilitates simultaneous exclusion of objects (e.g., excluding objects that are other than given objects and those objects which belong to the same category of given objects) so that one can comprehend the uniqueness of an object by excluding not only convergent but divergent objects (Thompson, 2021).

Our findings also reveal that there are some constraints to the applicability of Sikh mindfulness practices in certain circumstances. For instance, self-distanced mindful practices within Sikhism involve contemplation of past actions and experiences in order to identify the sources of negative impulses. However, in certain situations, that could lead to negative self-image and trigger avoidance behavior. The participants' experience also illustrate that compassion and empathy may not be successful in dealing with certain people. In such situations, the participants employed the Sikh ethos of detached balance and to navigate their moral dilemmas and gain insight on the situations in which compassion/empathy may not work. Another challenge in Sikh mindfulness may emerge when community-based mindfulness may be rooted in self-promotion rather than genuine community-based benefits. In such situations, our participants point to the need to use contemplation through *Simran* and subscribing to the tenets of righteousness thoughts and conduct to maintain the "other-oriented" focus in their mindful practices. This points out to how mindful practices within different traditions are intrinsically rooted in the spiritual principles and practices within the tradition and cannot be seen in isolation.

8.14 Conclusions

This study introduces dynamic mindfulness practices from the Buddhist and Sikh traditions that move beyond the focus on secular mindfulness approaches that emphasizes "moment" awareness (e.g., Purser & Milillo, 2015) and often considers mindfulness as a trait in the subject matter of business ethics (e.g., Gentina et al., 2020; Pandey et al., 2018; Reb et al., 2020; Small & Lew, 2021). Mindfulness practices in introduced in this chapter reflect a perspective-taking approach that highlights the importance of remembering moral philosophical and theological teachings of the traditions to guide other-focused ethical behaviors and decision-making in mindfulness practices that can promote sustainability approaches. We extend (1) the literature on mindfulness in business ethics (e.g., Kreplin et al., 2018; Orazi et al., 2021) by showing other-focused orientations in tradition-specific mindfulness practices that can navigate moral awareness in moral reasoning (Rest, 1986) and reject ego-centeredness of individual characteristics to acknowledge the importance of context-relevant approach (Trevino, 1986) to navigate individuals' moral judgment and reasoning and (2) the literature on mindfulness and environmental sustainability (Mueller & Greenwood, 2015; Sol & Wals, 2015) by demonstrating how ethical dimensions of mindfulness can motivate orientations toward sustainability through other-centeredness.

It would be insightful to study mindful practices from other spiritual/faith traditions in order to expand and enrich the emerging and complex discourse on mindful practices and its implications for organizational behavior. From a methodological perspective, our study demonstrates the utility of phenomenological research in developing holistic understanding of individual's mindful practices. It would be useful to see how phenomenological, qualitative research methodologies may be employed to other areas of research in order to explore the influence of mindful practices in areas such as ethical decision-making, pro-social behavior, and environmental conscious. Existing studies on mindful practices in these areas tend to be dominated by quantitative, survey-based methodologies. The use of qualitative approaches can facilitate in developing a more nuanced understanding of the influence of mindful practice, based on "thick descriptions." Another limitation of the study is that we did not categorize our participants according to mastery (years of mindful practice) and its implications for moral reasoning. It would be useful to conduct follow-up research on this subject and see how the contestations and limitations in mindful practice are navigated by more experienced practitioners versus those who are new to mindful practice.

References

Allen, T. D., & Kiburz, K. M. (2012). Trait mindfulness and work–family balance among working parents: The mediating effects of vitality and sleep quality. *Journal of Vocational Behavior, 80*(2), 372–379.

Anālayo. (2010). *Satipaṭṭhāna: The direct path to realization*. Windhorse Publications.

Baer, R., Crane, C., Miller, E., & Kuyken, W. (2019). Doing no harm in mindfulness-based programs: Conceptual issues and empirical findings. *Clinical Psychology Review, 71*, 101–114.

Barbaro, N., & Pickett, S. M. (2016). Mindfully green: Examining the effect of connectedness to nature on the relationship between mindfulness and engagement in pro-environmental behavior. *Personality and Individual Differences, 93*, 137–142.

Bernstein, A., Hadash, Y., Lichtash, Y., Tanay, G., Shepherd, K., & Fresco, D. M. (2015). Decentering and related constructs: A critical review and metacognitive processes model. *Perspectives on Psychological Science, 10*(5), 599–617.

Bierly, P. E., Kessler, E. H., & Christensen, E. W. (2000). Organizational learning, knowledge and wisdom. *Journal of Organizational Change Management, 13*(6), 595–618.

Bluth, K., Gaylord, S., Campo, R., Mullarkey, M. C., & Hobbs, L. (2016). Making friends with yourself: A mixed methods pilot study of a mindful self-compassion program for adolescents. *Mindfulness, 7*, 479–492.

Bodhi, B. (1994). *The noble eightfold path: The way to the end of suffering*. Buddhist Publication Society.

Bolderdijk, J. W., Steg, L., Geller, E. S., Lehman, P. K., & Postmes, T. (2013). Comparing the effectiveness of monetary versus moral motives in environmental campaigning. *Nature Climate Change, 3*(4), 413–416.

Brogaard, B. (2008). Moral contextualism and moral relativism. *The Philosophical Quarterly, 58*(232), 385–409.

Broome, J. (2008). The ethics of climate change. *Scientific American, 298*, 96–102.

Brown, K. W., & Kasser, T. (2005). Are psychological and ecological well-being compatible? The role of values, mindfulness, and lifestyle. *Social Indicators Research, 74*(2), 349–368.

Brown, K. W., & Ryan, R. M. (2004). *Perils and promise in defining and measuring mindfulness: Observations from experience*. Clinical Psychology: Science and Practice.

Chan, C., & Ananthram, S. (2019). Religion-based decision making in Indian multinationals: A multi-faith study of ethical virtues and mindsets. *Journal of Business Ethics, 156*(3), 651–677.

Chinn, P. W. (2015). Place and culture-based professional development: Cross-hybrid learning and the construction of ecological mindfulness. *Cultural Studies of Science Education, 10*(1), 121–134.

Clayton, B. (2009). Śāntideva, virtue, and consequentialism. In J. Powers & C. Prebish (Eds.), *Destroying Mara forever: Buddhist ethics essays in Honor of Damien Keown*. Snow Lion.

Cohen, L., Manion, L., & Morrison, K. (2018). *Research methods in education*. Routledge.

Cope, J. (2011). Entrepreneurial learning from failure: An interpretative phenomenological analysis. *Journal of Business Venturing, 26*(6), 604–623.

Corbin, J., & Strauss, A. (2008). *Basics of qualitative research: Techniques and procedures for developing grounded theory* (3rd ed.). Sage.

Craft, J. L. (2013). A review of the empirical ethical decision-making literature: 2004–2011. *Journal of Business Ethics, 117*(2), 221–259.

Creswell, J. W. (2013). *Qualitative inquiry and research design: Choosing among five approaches*. Sage.

Dane, E. (2011). Paying attention to mindfulness and its effects on task performance in the workplace. *Journal of Management, 37*(4), 997–1018.

Dane, E., & Brummel, B. J. (2014). Examining workplace mindfulness and its relations to job performance and turnover intention. *Human Relations, 67*(1), 105–128.

Dariotis, J., Mirabal-Beltran, R., Cluxton-Keller, F., Feagans Gould, L., Greenberg, M., & Mendelson, T. (2016). A qualitative evaluation of student learning and skills use in a school-based mindfulness and yoga program. *Mindfulness, 7*, 76–89.

Dhandra, T. K., & Park, H. J. (2018). Mindfulness and gender differences in ethical beliefs. *Social Responsibility Journal, 14*(7).

Donald, J. N., Sahdra, B. K., Van Zanden, B., Duineveld, J. J., Atkins, P. W., Marshall, S. L., & Ciarrochi, J. (2019). Does your mindfulness benefit others? A systematic review and meta-analysis of the link between mindfulness and prosocial behaviour. *British Journal of Psychology, 110*(1), 101–125.

Dunne, J. D. (2015). Buddhist styles of mindfulness: A heuristic approach. In *Handbook of mindfulness and self-regulation* (pp. 251–270). Springer.

Edwards, A. (2015). *The heart of sustainability: restoring ecological balance from the inside out*. New Society Publishers.

Epstein, M. (1988). The deconstruction of the self: Ego and "egolessness" in Buddhist insight meditation. *The Journal of Transpersonal Psychology, 20*(1), 61–69.

Ericson, T., Kjønstad, B. G., & Barstad, A. (2014). Mindfulness and sustainability. *Ecological Economics, 104*, 73–79.

Feinberg, M., & Willer, R. (2013). The moral roots of environmental attitudes. *Psychological Science, 24*, 56–62.

Feinberg, M., Kovacheff, C., Teper, R., & Inbar, Y. (2019). Understanding the process of moralization: How eating meat becomes a moral issue. *Journal of Personality and Social Psychology, 117*(1), 50.

Felver, J., & Jennings, P. (2017). Applications of mindfulness-based interventions in school settings: An introduction. *Mindfulness, 7*, 1–4.

Fischer, D., Stanszus, L., Geiger, S., Grossman, P., & Schrader, U. (2017). Mindfulness and sustainable consumption: A systematic literature review of research approaches and findings. *Journal of Cleaner Production, 162*, 544–558.

Foa, E. B., & Kozak, M. J. (1986). Emotional processing of fear: Exposure to corrective information. *Psychological Bulletin, 99*(1), 20.

Garrigan, B., Adlam, A. L., & Langdon, P. E. (2018). Moral decision-making and moral development: Toward an integrative framework. *Developmental Review, 49*, 80–100.

Geiger, S. M., Otto, S., & Schrader, U. (2018). Mindfully green and healthy: An indirect path from mindfulness to ecological behavior. *Frontiers in Psychology, 8*, 2306.

Gentina, E., Daniel, C., & Tang, T. L. P. (2020). Mindfulness reduces avaricious monetary attitudes and enhances ethical consumer beliefs: Mindfulness training, timing, and practicing matter. *Journal of Business Ethics, 173*, 301–323.

Gethin, R. (1998). *The foundations of Buddhism*. Oxford University Press.

Gilbert, P. (2013). *Mindful compassion: Using the power of mindfulness and compassion to transform our lives*. Hachette.

Gilbert, P. (2019). Explorations into the nature and function of compassion. *Current Opinion in Psychology, 28*, 108–114.

Gill, M. J. (2014). The possibilities of phenomenology for organizational research. *Organizational Research Methods, 17*(2), 118–137.

Good, D. J., Lyddy, C. J., Glomb, T. M., Bono, J. E., Brown, K. W., Duffy, M. K., et al. (2016). Contemplating mindfulness at work: An integrative review. *Journal of Management, 42*(1), 114–142.

Gould, S. J. (1995). The Buddhist perspective on business ethics: Experiential exercises for exploration and practice. *Journal of Business Ethics, 14*(1), 63–70.

Grossman, P. (2015). Mindfulness: awareness informed by an embodied ethic. *Mindfulness, 6*(1), 17–22.

Hafenbrack, A. C., Cameron, L. D., Spreitzer, G. M., Zhang, C., Noval, L. J., & Shaffakat, S. (2020). Helping people by being in the present: Mindfulness increases prosocial behavior. *Organizational Behavior and Human Decision Processes, 159*, 21–38.

Hofmann, W., Meindl, P., Mooijman, M., & Graham, J. (2018). Morality and self-control: How they are intertwined and where they differ. *Current Directions in Psychological Science, 27*, 286–291.

Hurst, K., & Stern, M. J. (2020). Messaging for environmental action: The role of moral framing and message source. *Journal of Environmental Psychology, 68*, 101394.

Jamieson, D. (2010). Climate change, responsibility, and justice. *Science and Engineering Ethics, 16*, 431–445.

Jia, F., Soucie, K., Alisat, S., Curtin, D., & Pratt, M. (2017). Are environmental issues moral issues? Moral identity in relation to protecting the natural world. *Journal of Environmental Psychology, 52*, 104–113.

Jinpa, T. (2019). The question of mindfulness' connection with ethics and compassion. *Current Opinion in Psychology, 28*, 71–75.

Jones, T. M. (1991). Ethical decision making by individuals in organizations: An issue-contingent model. *Academy of Management Review, 16*(2), 366–395.

Kabat-Zinn, J. (2019). Foreword: Seeds of a necessary global renaissance in the making: The refining of psychology's understanding of the nature of mind, self, and embodiment through the lens of mindfulness and its origins at a key inflection point for the species. *Current Opinion in Psychology, 28*, xi–xvii.

Kalafatoğlu, Y., & Turgut, T. (2017). Another benefit of mindfulness: Ethical behaviour. *International Journal of Social Sciences and Education Research, 3*(3), 772–782.

Kalra, G., Bhui, K., & Bhugra, D. (2013). Does Guru Granth Sahib describe depression? *Indian Journal of Psychiatry, 55*(6), 195–200.

Kang, C., & Whittingham, K. (2010). Mindfulness: A dialogue between Buddhism and clinical psychology. *Mindfulness, 1*(3), 161–173.

Kasser, T., & Sheldon, K. M. (2009). Time affluence as a path toward personal happiness and ethical business practice: Empirical evidence from four studies. *Journal of Business Ethics, 84*(2), 243–255.

Kern, M. C., & Chugh, D. (2009). Bounded ethicality: The perils of loss framing. *Psychological Science, 20*(3), 378–384.

Krägeloh, C. U. (2016). Importance of morality in mindfulness practice. *Counseling and Values, 61*(1), 97–110.

Kreplin, U., Farias, M., & Brazil, I. A. (2018). The limited prosocial effects of meditation: A systematic review and meta-analysis. *Scientific Reports, 8*(2403), 1–10.

Kross, E., & Ayduk, O. (2011). Making meaning out of negative experiences by self-distancing. *Current Directions in Psychological Science, 20*(3), 187–191.

Kross, E., Ayduk, O., & Mischel, W. (2005). When asking "why" does not hurt distinguishing rumination from reflective processing of negative emotions. *Psychological Science, 16*(9), 709–715.

Kudesia, R. S., & Nyima, V. T. (2015). Mindfulness contextualized: An integration of Buddhist and neuropsychological approaches to cognition. *Mindfulness, 6*(4), 910–925.

Lampe, M., & Engleman-Lampe, C. (2012). Mindfulness-based business ethics education. *Academy of Educational Leadership Journal, 16*(3), 99.

Langer, E. J., & Moldoveanu, M. (2000). Mindfulness research and the future. *Journal of Social Issues, 56*(1), 129–139.

Lau, M. A., Bishop, S. R., Segal, Z. V., Buis, T., Anderson, N. D., Carlson, L., et al. (2006). The Toronto mindfulness scale: Development and validation. *Journal of Clinical Psychology, 62*(12), 1445–1467.

Lehnert, K., Park, Y. H., & Singh, N. (2015). Research note and review of the empirical ethical decision-making literature: Boundary conditions and extensions. *Journal of Business Ethics, 129*(1), 195–219.

Libby, L. K., & Eibach, R. P. (2002). Looking back in time: Self-concept change affects visual perspective in autobiographical memory. *Journal of Personality and Social Psychology, 82*(2), 167.

Lincoln, Y. S., & Guba, E. G. (1985). *Naturalistic inquiry*. Sage.

Magid, B., & Poirier, M. R. (2016). The three shaky pillars of Western Buddhism: Deracination, secularization, and instrumentalization. In *What's wrong with mindfulness (and what isn't)* (pp. 39–52).

Mandair, A. P. S. (2012). *Encyclopaedia of Indian religions: Sikhism*. Springer.

Markowitz, E. M., & Shariff, A. F. (2012). Climate change and moral judgement. *Nature Climate Change, 2*, 243–247.

McLoughlin, N., Corner, A., Clarke, J., Whitmarsh, L., Capstick, S., & Nash, N. (2019). *Mainstreaming low-carbon lifestyles*. Climate Outreach.

Mihelič, K. K., & Culiberg, B. (2019). Reaping the fruits of another's labor: The role of moral meaningfulness, mindfulness, and motivation in social loafing. *Journal of Business Ethics, 160*(3), 713–727.

Monteiro, L. M., Musten, R. F., & Compson, J. (2015). Traditional and contemporary mindfulness: Finding the middle path in the tangle of concerns. *Mindfulness, 6*(1), 1–13.

Mooijman, M., Meindl, P., Meindl, P., Oyserman, D., Monterosso, J., Dehghani, M., et al. (2018). Resisting temptation for the good of the group: Binding moral values and the moralization of self-control. *Journal of Personality and Social Psychology, 115*, 585–599.

Moustakas, C. (1994). *Phenomenological research methods*. Sage.

Mueller, M. P., & Greenwood, D. A. (2015). Ecological mindfulness and cross-hybrid learning: a special issue. *Cultural Studies of Science Education, 10*(1), 1–4.

Neff, K. D., & Dahm, K. A. (2015). Self-compassion: What it is, what it does, and how it relates to mindfulness. In *Handbook of mindfulness and self-regulation* (pp. 121–137). Springer.

Nesbitt, E. (2005). *Sikhism: A very short introduction*. Oxford University Press.

Nielsen, K. S., & Hofmann, W. (2021). Motivating sustainability through morality: A daily diary study on the link between moral self-control and clothing consumption. *Journal of Environmental Psychology, 73*, 101551.

Nonaka, I., Chia, R., Holt, R., & Peltokorpi, V. (2014). Wisdom, management and organization. *Opinion in Psychology, 28*, 71–75.

Orazi, D. C., Chen, J., & Chan, E. Y. (2021). To erect temples to virtue: Effects of state mindfulness on other-focused ethical behaviors. *Journal of Business Ethics, 169*(4), 785–798.

Pace, S. (2013). Does religion affect the materialism of consumers? An empirical investigation of Buddhist ethics and the resistance of the self. *Journal of Business Ethics, 112*(1), 25–46.

Pandey, A., Chandwani, R., & Navare, A. (2018). How can mindfulness enhance moral reasoning? An examination using business school students. *Business Ethics: A European Review, 27*(1), 56–71.

Panesar, D. S. (2017). Meditation. In A. S. Mandair (Ed.), *Sikhism: Encyclopaedia of Indian Religions*. Springer.
Petrova, K., Nevarez, M. D., Waldinger, R. J., Preacher, K. J., & Schulz, M. S. (2021). Self-distancing and avoidance mediate the links between trait mindfulness and responses to emotional challenges. *Mindfulness, 12*, 947–958.
Poore, J., & Nemecek, T. (2018). Reducing food's environmental impacts through producers and consumers. *Science, 992*, 987–992.
Poortinga, W., Whitmarsh, L., Steg, L., Bohm, G., & Fisher, S. (2019). Climate change perceptions and their individual-level determinants: A cross- European analysis. *Global Environmental Change, 55*, 25–35.
Purser, R. E., & Milillo, J. (2015). Mindfulness revisited: A Buddhist-based conceptualization. *Journal of Management Inquiry, 24*(1), 3–24.
Pye, O. (2019). Commodifying sustainability: Development, nature and politics in the palm industry. *World Development, 121*, 218–228.
Qiu, J. X., & Rooney, D. (2019). Addressing unintended ethical challenges of workplace mindfulness: A four-stage mindfulness development model. *Journal of Business Ethics, 157*(3), 715–730.
Reb, J., & Atkins, P. W. (Eds.). (2015). *Mindfulness in organizations: Foundations, research, and applications*. Cambridge University Press.
Reb, J., Allen, T., & Vogus, T. J. (2020). *Mindfulness arrives at work: Deepening our understanding of mindfulness in organizations. Organizational Behavior and Human Decision Processes, 159*, 1–7.
Reed, D. C. (2009). A multi-level model of moral functioning revisited. *Journal of Moral Education, 38*(3), 299–313.
Rest, J. R. (1986). *Moral development: Advances in research and theory*. Praeger.
Robinson, J. A., & Swanson, K. L. (1993). Field and observer modes of remembering. *Memory, 1*(3), 169–184.
Robinson, J., Sinclair, M., Tobias, J., & Choi, E. (2017). More dynamic than you think: Hidden aspects of decision-making. *Administrative Sciences, 7*(3), 1–29.
Roxas, B. (2021). Environmental sustainability engagement of firms: The roles of social capital, resources, and managerial entrepreneurial orientation of small and medium enterprises in Vietnam. *Business Strategy and the Environment, 30*(4), 2194–2208.
Rozin, P. (1999). The process of moralization. *Psychological Science, 10*, 218–221.
Ruedy, N. E., & Schweitzer, M. E. (2010). In the moment: The effect of mindfulness on ethical decision making. *Journal of Business Ethics, 95*(1), 73–87.
Sahdra, B. K., Shaver, P. R., & Brown, K. W. (2010). A scale to measure nonattachment: A Buddhist complement to Western research on attachment and adaptive functioning. *Journal of Personality Assessment, 92*(2), 116–127.
Sahota, G. S. (2011). Guru Nanak and rational civil theology. *Sikh Formations, 7*(2), 131–143.
Scott, J. (1997). Corporate business and capitalist classes. Oxford University Press, Oxford.
Sevinc, G., & Lazar, S. W. (2019). How does mindfulness training improve moral cognition: a theoretical and experimental framework for the study of embodied ethics. *Current Opinion in Psychology, 28*, 268–272.
SGGS. (n.d.). *Sri Guru Granth Sahib in English translation*. Publication Bureau, Punjabi University.
Shackle, C., Singh, G., & Mandair, A. (2001). *Sikh religion, culture and ethnicity*. Curzon Press.
Shapiro, S. L., Jazaieri, H., & Goldin, P. R. (2012). Mindfulness-based stress reduction effects on moral reasoning and decision making. *The Journal of Positive Psychology, 7*(6), 504–515.
Singh, K. (2004). *A history of the Sikhs: Volume 1: 1469-1839* (2nd ed.). Oxford University Press.
Singh, N. G. K. (2005). *The birth of the Khalsa, a feminist RE-memory of Sikh Identity*. State University of New York Press.
Singh, N. (2020). Sikhism and sustainability. *International Journal of Sikh Studies, 2*(1), 66–85.

Singh, P., & Singh, N. (2019). Political economy of bioenergy transitions in developing countries: A case study of Punjab, India. *World Development, 124*(2019), 104360.

Singer, T., & Klimecki, O. M. (2014). Empathy and compassion. *Current Biology, 24*(18), R875–R878.

Small, C., & Lew, C. (2021). Mindfulness, moral reasoning and responsibility: Towards virtue in ethical decision-making. *Journal of Business Ethics, 169*(1), 103–117.

Smith, J. A., Flowers, P., & Larkin, M. (2009). *Interpretative phenomenological analysis: Theory, method and research.* Sage.

Sol, J., & Wals, A. E. (2015). Strengthening ecological mindfulness through hybrid learning in vital coalitions. *Cultural Studies of Science Education, 10*(1), 203–214.

Sosik, J. J., Chun, J. U., Ete, Z., Arenas, F. J., & Scherer, J. A. (2019). Self-control puts character into action: Examining how leader character strengths and ethical leadership relate to leader outcomes. *Journal of Business Ethics, 160*, 765–781.

Sutcliffe, K. M., Vogus, T. J., & Dane, E. (2016). Mindfulness in organizations: A cross-level review. *Annual Review of Organizational Psychology and Organizational Behavior, 3*, 55–81.

Tenbrunsel, A. E., & Messick, D. M. (2004). Ethical fading: The role of self-deception in unethical behavior. *Social justice research, 17*(2), 223–236.

Thich, N. H. (1999). *The miracle of mindfulness*. Beacon.

Thompson, E. (2021). What's in a concept? Conceptualizing the nonconceptual in Buddhist Philosophy and cognitive Science. In Coseru, C. (Ed), Reasons and Empty Persons: Mind, Metaphysics, and Morality: Essays in Honor of Mark Siderits, pp. 165–210. London: Springer.

Tran, V. V., Park, D., & Lee, Y. C. (2020). Indoor air pollution, related human diseases, and recent trends in the control and improvement of indoor air quality. *International Journal of Environmental Research and Public Health, 17*(8), 2927.

Trevino, L. K. (1986). Ethical decision making in organizations: A person-situation interactionist model. *Academy of Management Review, 11*(3), 601–617.

Tsoukas, H., & Cummings, S. (1997). Marginalization and recovery: The emergence of Aristotelian themes in organization studies. *Organization studies, 18*(4), 655–683.

Van Dam, N. T., Sheppard, S. C., Forsyth, J. P., & Earleywine, M. (2011). Self-compassion is a better predictor than mindfulness of symptom severity and quality of life in mixed anxiety and depression. *Journal of Anxiety Disorders, 25*(1), 123–130.

Verhaeghen, P. (2015). Good and well: The case for secular Buddhist ethics. *Contemporary Buddhism, 16*(1), 43–54.

Vokey, D. (2001). *Moral discourse in a pluralistic world.* University of Notre Dame Press.

Vu, M. C., & Gill, R. (2018). Is there corporate mindfulness? An exploratory study of Buddhist-enacted spiritual leaders' perspectives and practices. *Journal of Management, Spirituality & Religion, 15*(2), 155–177.

Vu, M. C., & Tran, T. (2021). Trust issues and engaged Buddhism: The triggers for skillful managerial approaches. *Journal of Business Ethics, 169*(1), 77–102.

Vu, M. C., Wolfgramm, R., & Spiller, C. (2018). Minding less: Exploring mindfulness and mindlessness in organizations through skillful means. *Management Learning, 49*(5), 578–594.

Wells, A. (2005). Detached mindfulness in cognitive therapy: A metacognitive analysis and ten techniques. *Journal of Rational-Emotive and Cognitive-Behavior Therapy, 23*(4), 337–355.

Wolsko, C. (2017). Expanding the range of environmental values: Political orientation, moral foundations, and the common in group. *Journal of Environmental Psychology, 51*, 284–294.

Part VI
Abrahamic Religions

Chapter 9
Judaism and Ecological Discourse: What a Jewish Religious Perspective Offers to Contemporary Sustainability Dialogue

Susan S. Case

Abstract This chapter explores ways classical Jewish sources, Jewish law (Halakha), and the Hebrew Bible (Tanakh) can be analyzed around ecological themes contributing to global sustainability practices to protect our environment. Environmental motifs are found throughout these texts. Perspectives used focus on interconnecting humanity and nature linking environmental stewardship, protection, and the idea of *tikkun olam*, repairing the world. This Jewish religious perspective, with its rich history of respect for nature and non-human life, interrelated with social justice, offers insight into "just sustainability," serving as a model and guide for a more sustainable future. A key Jewish contribution to environmental ethics is interdependence among God, human needs, and claims of nature, with a responsibility to use nature wisely and responsibly, preserving it for future generations and caring for its people justly. We explore modern environmental concerns through this biblical lens including stewardship, climate change, resource conservation, urban planning, sensitivity to animal welfare, and environmental health. Incorporating Judaism's ancient wisdom from more than 2500 years ago, with its balanced approach to environmental issues, can inspire positive individual and organizational actions enhancing ecological discourse.

Keywords Judaism and ecology · Hebrew scriptures and environmental ethics · Environmental stewardship · "Just sustainability" · *tikkun olam* · Environmental justice

S. S. Case (✉)
Case Western Reserve University, Cleveland, OH, USA
e-mail: susan.case@case.edu

N. Singh et al. (eds.), *Faith Traditions and Sustainability*, Management, Change, Strategy and Positive Leadership, https://doi.org/10.1007/978-3-031-41245-5_9

9.1 Introduction

A defining issue of the twenty-first century is the environment, especially the impact of human actions on ecosystems (Weyler, 2018; United Nations Department of Global Communications, 2020). Various faiths acknowledge the need for environmental stewardship with their respective holy texts urging adherents to be caretakers of the earth and its biodiversity (United Nations Environmental Programme, 2019). Biblical, Rabbinic, and Jewish sources have addressed similar issues for at least 2500 years. There are 613 commandments in the Torah (*mitzvah* or duties), which provide religious and moral guidance for Jewish conduct. Many have to do with how we are to plant, plow, harvest, earn, spend, eat, dispose of waste, care for others, and live in a community in a way God finds acceptable (Weintraub, 2022). These ethical principles for behavior instruct people on ways to sustain the earth and God's creations. The Torah teaches that God created the world for the benefit of mankind with both rights and responsibility. The world belongs to God, but we are its stewards, charged with the task of working it and protecting it (Bereishit 2:15). Through the prism of the Torah's moral and legal guidelines is a key contribution of Judaism to an environmental ethic of interdependence amongst God, people, and nature, with an obligation to not only use nature wisely and responsibly, preserving it for future generations, but to also care for others justly. Judaism provides inspiration and a code of ways to live on earth with practical benefits which can guide our thinking about today's ecological issues (Bernstein, 2018).

The sacred texts of Judaism, called the Jewish Canon (Lim, 2013), provide a source of moral norms and beliefs with a major emphasis on correcting the world's imperfections.[1] In this chapter, I draw on this Canon including commandments and ethical principles in the Torah for sustaining the earth and God's creations. Classical Jewish texts express a philosophically well-grounded environmental ethic about the relationship of nature and man (Tirosh-Samuelson, 2001).[2] They provide an inseparable link with religious practices consistent with environmental protection and serving our fellow human beings. They have much to teach about caring for world resources and environmental issues even though initially developed for an agrarian society. Judaism stresses preservation and creation of natural resources for

[1] The author grew up in Reform and Reconstructionist Jewish communities with Jewish educator parents, all stressing moral behaviors required to live a Jewish life, making a difference in the world.

[2] Exploring these ecological perspectives requires an understanding of the place of the Hebrew Bible in Jewish thought and practice. Unlike the Christian Bible, biblical Jewish texts are not stand-alone sources. You read them through a lens of rabbinic texts with later commentaries and interpretations developed after the Roman destruction of the Jerusalem Temple in 70 CE (Belser, 2022). The Mishnah (canonized in 200 CE) is one of the earliest texts of rabbinic law and practice. The Babylonian Talmud is a commentary on the Mishnah including debates between rabbis about Jewish law and practice. Judaism is not settled beliefs and law. Rather, in Jewish tradition, the Talmud emphasizes a process of inquiry and argumentation (Belser, 2022, p. 2). It is this Midrashic tradition that brings contemporary ecological concerns into Judaism. Rabbis employ commandments for each age in which they lived, attending to their world, fashioning everyday life around these commandments (Kornfeld, 2021, p. 1014).

future generations, a kind of intergenerational justice. Serving God and serving our fellow human beings links moral consciousness and environmentalism in the modern Jewish environmental movement drawing on Jewish sources (Bolton-Fasman, 2020; Jackson, 2022; Neril & Dee, 2020, 2021). The bridge between these is "just sustainability," an attempt to distribute natural resources and environmental burdens equitably, while simultaneously creating a long-term biodiverse ecologically sustainable world (Jacobs, 2009, pp. 179–180).

A Talmudic story of the sage Choni illustrates that the act of preserving natural resources is an act of faith, concerned both with nature and humanity. While walking along a road, he observed a 100-year-old Jewish man planting a fig tree in his garden. Choni asked, "How long will it take you for this tree to bear fruit?" "Seventy years," the man replied. Choni then asked if he expected to live that long to enjoy the fruits of his labor. He replied: "I was born into a world flourishing with ready pleasures. My ancestors planted for me and now I plant for my children…" (Babylonian Talmud Ta'anit 23b). Because God makes us think of eternity, like in the story of Choni, faith is a powerful counterforce to self-interest and short-term gain (Rousseau, 1762/1968) empowering the exercising of freedom for the common good and the ethic of responsibility to posterity (Case & Smith, 2013).

9.2 Context for Ecological Discourse

Humans contribute to global warming through pollution and ozone layer depletion, introducing harmful substances into the environment by acid rain, oil spills, toxic chemicals, improper waste disposal, and pesticides. This results in more frequent extreme weather events involving intense heat, drought, wildfires, rainfall, hurricanes, cyclones, flooding, and soil erosion (United Nations, 2022) increasingly damaging agriculture, the wilderness, and animals. Every region is experiencing these extremes with warming melting ice caps and glaciers, leading to sea level rising 30–60 centimeters by 2100 (United Nations, 2022). One impact of this rise is salt-contaminated soil, making fields useless (Chen & Mueller, 2018). By 2030, more intense storms involving flooding will increase by 40%, averaging 596 extreme events a year, as will more frequent droughts across one-third of the globe leading to famine displacing 700 million people (United Nations, 2022). In 2020, in the United States, crop damage and rangeland loss of over $6.5 billion resulted from natural disasters (Munch, 2021). Ten percent of the world's wilderness has been lost since the early 1990s (Aycrigg et al., 2022; Watson et al., 2016). Over 450,000 animal species are at risk of extinction over the coming decade (United Nations, 2022).

Whether by filling in wetlands, overfishing, or hunting endangered species, people destroy natural habitats. Valleys flood through building dams (Stokstad, 2022). Infrastructure expansion leads to erosion (Borrelli et al., 2017). We have a biodiversity crisis through altered interactions of plants and animals important to ecosystem flourishing (Shao, 2022). Habitats are decimated by housing development, mining, agriculture, and killing animals for sport, use, and food. Ninety percent of the world's fishermen cannot earn a livelihood because fish have disappeared due to ocean warming and acidification (United Nations, 2022). In the last 50 years, human beings have destroyed 50% of all forests and woodland for timber, construction, fuel, and growing crops with four commodities (beef, soy, palm oil, and wood products) driving most global deforestation (Lock, 2021; Sustainability and unsustainability, n.d.; Union of Concerned Scientists, 2016). Most species need oxygen, provided by trees to survive. The carbon dioxide resulting from deforestation has contributed to a greenhouse effect. Urgent action is needed if we are to save the planet (Sustainability & climate change, 2022).

On a microlevel, the inequitable and disproportionately heavy exposure of poor, minority, and disenfranchised populations to air pollution (Tessum et al., 2021) and toxic chemical contamination of land and water through the accumulation of chemical, heavy metal, biological and nuclear waste has led to disease, disability, and death of people of all ages (Andersen et al., 2021; Landrigan et al., 2010). Cancer is the second leading cause of death in American children; asthma has doubled in frequency since the 1980s resulting in the most pediatric hospitalizations and highest school absenteeism; and birth defects are the leading cause of infant death (Landrigan et al., 2010, p. 178). Air pollution and contaminated land and water are responsible for one in six deaths worldwide (Landrigan et al., 2018). Environmentally caused diseases like cancer and immune-system problems have increased across the globe (Andersen et al., 2021). 4.2 million deaths in 2019 were attributed to air pollution from traffic, industry, power generators, waste burning, and residential fuel consumption (United Nations, 2022). 829,000 people die yearly from diseases directly attributable to unsafe water and inadequate sanitation and waste removal (United Nations, 2022). By 2030, it is estimated that more than 2.8 billion people will be without accessible water or adequate sanitation (United Nations, 2022).

Our environmental challenge is combining environmental conservation issues like protecting endangered species, fighting climate change, and reducing the number of resources used by human activity with environmentalists working on political social justice issues like waste disposal, the location of polluting chemical plants, and regulating harmful chemicals and greenhouse gas emissions. Most ecological discourse is done through a binary model separating the concern for nature with a concern for humanity.

The Jewish perspective, illustrated through the story of Choni, contrasts with Western economic views (Friedman, 1962; Smith, 1937) where nature exists to be exploited. Financially unremunerative activities, even those essential to the well-being of society, are not valued (Dhar & Case, 2023) like giving life and caring for life, supporting human development. They are viewed as "reproductive" labor, not part of a "productive" economy (Eisler, 2007, p. 148; 2017b). Other economies not focused on self-interest, like household, natural, and unpaid community volunteer economies, are left out of this market economic system (Eisler, 2007, 2017b). Ancient scriptural teaching embeds moral responsibilities for creating equitable, sustainable, and caring systems (Dhar & Case, 2023) moving us beyond capitalism as it is currently practiced. Judaism imposes restrictions on the ways and extent to which people can use the natural environment. Ecological action is a Jewish obligation (Belser, 2022; Vogel, 2001).

9.2.1 Responsible Interdependence: People, Land, and God

Medieval Jewish thinkers believed there were 613 *mitzvot* in the Torah, creating key obligations and prohibitions of Jewish law for behavior. The obligations to keep these is at the core of Jewish religious practice (Belser, 2022; Brauner, 2011). Unlike many religions primarily focused on beliefs, Judaism is centered around behavior.

Environmental concerns involving preserving, protecting, and nourishing our planet are scattered throughout rabbinic texts now redeployed as Jewish obligations (Belser, 2022). Since Jews view their role on earth as God's co-partners in the upkeep of the world, there are important social justice and communal responsibilities for the betterment of society. There is an obligation to work for *tikkun olam*, "the repair of the world" with expectations to mend and improve it (Rose et al., 2008). The burdens of pollution, toxic waste, and poisoned resources are not equally distributed across society with marginal communities throughout the world disproportionately impacted by climate change (Calil et al., 2021; Patnaik et al., 2020; Islam & Winkel, 2017). They are least able to prepare for, and recover from, heat waves, poor air quality, and flooding and are more likely to die of environmental causes (EPA, 2021; Yip, 2021). Repairing the world includes acting ethically and fairly to improve the quality of life for everyone (Rose et al., 2008).

The Hebrew Bible begins with Genesis, telling the story of God's creation of the world. In this book, the Jewish view of humanity and humans' relationship with God explains how we should relate and care for the world God has given us. The chapter describes creation of various animal species, climaxing with the creation of humankind establishing their dominion over the rest of the world (Genesis 1:28–29). On the seventh day, he wanted stillness in the world, making that day holy, the Sabbath (Sacks, 2005). The Jewish view of nature, reflected in the mentioned story of creation, is God created it for the sake of mankind who is to use God's creation, but man has a responsibility to use the land wisely and responsibly, protecting the world and its resources (Genesis 2:15). By thinking of the world as God's world and we as God's creatures, we become stewards of an existing universe needing sustaining for life after us (Berrett, 2011, A11).

Sacred Jewish texts are constructed in terms of responsibilities, not rights, with a major responsibility to correct the world's imperfections. Concerns for humanity and nature are inseparable with the health of the natural world and humanity intertwined. Contemporary Jewish environmental thinkers juxtapose stories that temper dominion and mastery with obligation for responsible stewardship (Belser, 2022; Regenstein, 2008; Sacks, 2016). Rabbinic midrash emphasizes restraint and responsibility for a balanced ecosystem acknowledging the intrinsic value of the natural world and everything in it (Belser, 2022, p. 7). The alignment of the two books of

Genesis recognize kinship between humans, animals, and our planet, requiring strict responsibility to care of the earth and its creatures (Kornfeld, 2021, p. 1023).

Judaism, with its God-centric focus, has a reverence for land not as an entitlement but as a responsibility, with its primary interest in the interdependent ecosystem of people, land, and God. This is a central theme in the entire Torah (Bernstein, 2018). Jewish environmental activism today has centered actions on approaches to Jewish practice that create a healthy and just future.[3] Linking ecological justice to social justice's protection of human life is the foundation of doing justice work Jewishly (Dobb, 2018; Tirosh-Samuelson, 2022).

9.2.2 Humanity and Nature in Jewish Tradition

The Jewish view of nature contrasts with currently accepted Western economic views developed by Adam Smith, who stated that work is a means of fulfilling one's self-interest and driving progress (Eisler, 2017a; Friedman, 1962; Smith, 1937). Ignored is the importance of nature's life-sustaining activities. Nature is viewed as an economic resource, economic growth is an end in itself; individual rights emphasize autonomy trumping responsibility to the collective; material consumption is desirable; government regulation should be limited impacting freedom, equality, and justice; and science is viewed with skepticism (Laszlo et al., 2020, pp. 4–6).

Jewish ethicists believe self-interest is to be overcome by the common good (Case & Smith, 2013). An ethic of interdependence of humans and nature requires respect for the earth and its people and environmental protection for plants, animals, the air, water, and soil (Case & Smith, 2013; Hoffman & Sandelands, 2005). Inequality, environmental degradation, global warming, and species extinction are intractable problems that need addressing (Case & Chavez, 2017; Dhar & Case, 2023).

Four attitudes distinguish Judaism's approach to humanity and nature from unrestrained capitalism which contributes to environmental degradation and global injustice. First, since Jews were slaves to Pharaoh in Egypt, they know unfair treatment, so they act empathetically, treating others kindly and justly (Deuteronomy 6:21). Oppression in any form is unethical (Exodus 22:20–26) with implications for ensuring everyone's welfare across the globe, focusing on justice, a Jewish value. Second, "The earth and all that is in it belong to God" (Psalms 24:1), makes people caretakers in the world who cannot abuse the Owner's property (Leviticus 25:23). The Torah, an environmental trailblazer, teaches the fragile balance of our

[3] There are many Jewish organizations across the Orthodox to Reform spectrum that do Jewish environmental activism. These include Shalom Center that promotes an activist ecological understanding of Judaism, Coalition of the Environment and Jewish Life (COEJL), Shomrie Adamah (Keepers of the Earth), Canfei Nesharim, Teva Learning Alliance, Heschel Center for Environmental Learning and Leadership (Israel), Jewish Climate Initiative and others (Dobb, 2018: Krone, June 23, 2021).

ecosystem that needs maintaining. No one should take more than they need from it. The lives and futures of people are interdependent with nature that is both to be used and preserved. There is a whole program in Torah for creating balanced distributions of resources across society (Exodus 22:24–26; Leviticus 25:36–37; Deuteronomy 23:20–21, 24:6, 10–13, 17) and commandments for sustainability. Judaism expects the world's wealth should be used on behalf of all inhabitants. In a world with ample food, allowing any people to go hungry violates Jewish stewardship (Case & Smith, 2013, p. 42). Third, since Genesis 1.27 states, "God created the human being in His image," people are expected to do right (Deuteronomy, 6:18), elevating compassion into moral social action. Humans have God-like capacities of power, consciousness, relationship, will, freedom, and life, which we are to exercise as wise stewards (Troster, 2011). Finally, sages spoke of going beyond the requirements of the law to enforce its spirit, practicing righteousness, and doing justice in human relationships (Case & Smith, 2013). These attitudes lead to balancing resource use, with people viewed as stewards of the world called to sustainably use the land that God created.

9.2.3 Jewish Perspectives on Ecological Protection

The obligation of people to respect and protect the environment appears throughout the Bible. The Torah recognizes nature as malevolent and beneficent.

9.2.4 Nature's Destructiveness

The story of the flood and Noah's life include lessons about how our actions affect the planet and its people. They tell what happens to civilizations when individuals abuse their land and each other. The Midrash says God gave humans 120 years to improve their corrupt ways of living with each other; but their behavior did not change. Noah, a good man, had demonstrated environmental concerns in how he lived his life. He was chosen by God to build an arc to save all species from an impending flood that would wipe out life on the planet. He was to take two of every species into an ark, committed to the preservation of species created in Genesis. Simultaneously, God created an ecological catastrophe destroying millions of plants, animals, and people (Genesis 6:12; Genesis 7:21–23), following it with a rainbow to assure such a natural catastrophe would never occur again. The rainbow reminds us we can take responsibility for the impact of our lives on others (Neril, 2014). This example combining the individual with responsibility for the collective benefit is a gift Judaism gave to the world (Cohn & Hill, 2007; Sacks, 2005). By emulating Noah, we can take responsibility for actions that sustain our planet.

Other biblical narratives also show nature as a life-threatening force. The story of Joseph includes 7 years of misery as Jews were forced to leave Egypt in search of food because the entire region was made barren. A critical portion of the Exodus narrative includes plagues foisted on the Egyptians, half including frogs, vermin, wild beasts, hail, and locusts, using natural forces to make their lives miserable. The Egyptian army succumbs to nature's destructiveness by drowning attempting to recapture fleeing Jews. In Deuteronomy 27.1, God tells Moses that if the Jewish people do not follow his commandments, natural catastrophes will follow. In the context of Jewish tradition, these God-created natural disasters serve divine purposes. Judaism appreciates nature, but nature is not to be worshipped as it was in pantheistic traditions of other ancient religions.

Like the people during the time of Noah, there are parallels to modern times. The natural world posts challenges to human survival. Since the 1970s, scientists have theorized that the warming of our planet is due to human consumption of fossil fuels (Maslin, 2004). The Intergovernmental Panel on Climate Change was created as a United Nations effort in 1988 to provide world governments explicit scientific views of what is happening to the world's climate (IPCC, 2022). The warnings have been mostly unheeded. Significant action has not been taken. The story of Noah reminds us of the critical need to bring about ecological balance.

As in Noah's time, many of the problems we face today include the destruction of wildlife and habitat as well as pollution of our food, air, and water. The Jewish approach to nature is distinctive because it links human treatment of God's earth and social justice. In Judaism, those who own land have a moral and religious obligation to support those who do not. In the agrarian times of the Torah's creation, the cultivated field was the primary model for the created universe. Parts of the land's produce – the corner of the field (*peah*), the gleanings of stalks (*leket*), the forgotten sheaf (*shikhekhah*), the separated fruits (*peret*), and the defective clusters (*olelot*) were to be given to those least advantaged who did not own land: the poor, the widow, the alien resident, and the Levite. In the Torah portion *Parashat Ki Tavo* (Deuteronomy 26:1–29:8), Jews are required to take a tenth of the yield and give it to those who are needy. This ensures that they are taken care of by their community. Treating others in society justly, protecting the dignity and sanctity of their lives, was tied to acts extended toward the land. This biblical principle provides religious justification for contemporary Jewish social ecology (Tirosh-Samuelson, 2001, p. 110).

9.2.5 Stewardship

As described earlier, the tradition of stewardship comes from Genesis where we are to partner in the continuing work of Creation. In the Garden of Eden, God charged Adam and Eve to tend the earth, teaching us to care for our planet to preserve that which God has created. Jews are encouraged *l'vadah ul'shamrah*, "to till and to tend" (Kohelet Rabbah 7:13), invoking the principle of stewardship. The Bible makes clear that the earth does not belong to us. It is not for conquest, ownership, nor domination. We do not have the right to do with earth as we choose (Genesis

1–2, Psalm 24:1, 1 Chron 29:10–16). The Biblical view is as follows: "The earth is the Lord's and the fullness thereof," with God claiming ownership and all property rights contingent on the degree we live in accordance with God's will (Lerner, 1994). God provides for humans, and we have a sacred duty to serve and guard nature, taking care of what we have been given. The Jewish theocentric worldview cultivates greater responsibility to future generations and respect for others on this world. This is stewardship in the Jewish religion.

Natural symbolism is common in Jewish tradition. The Torah is viewed as "a tree of life" (Proverbs 3:18). The Creation story is central throughout the Torah with the halakhic command of *bal tashchit*, "do not destroy," forbidding the cutting down of fruit trees during wars (Deuteronomy 20:19–20), the basis of an ethic of environmental responsibility (Sacks, 2016). The Talmud expanded the concept to include all forms of senseless damage and waste both of nature and by human beings including food, energy, or water (Vogel, 2001). A powerful Midrash states: "Take care not to spoil or destroy My world, for if you do, there will be nobody after you to make it right again" (Midrash Ecclesiastes Rabba 7:13). We have a responsibility to not spoil it for when we are wasteful, we violate our mandate to use Creation only for legitimate benefit considering its importance to future generations and others in the world (Troster, 2011). Jewish legal scholars apply *bal tashchit* to human needs, protecting economic interests. Only the wanton destructiveness of nature is wrong.

Environmental stewardship in modern sustainability dialogue draws on this biblical view. It is an action-oriented framework fostering social-ecological sustainability of a rapidly changing planet needing responsible use and protection of its natural environment through conservation and sustainability practices enhancing ecosystem resilience and human well-being (Chapin et al., 2010). In general, this stewardship focus is on transformative conservation of ecosystems to counter biodiversity loss and climate change (Fougères et al., 2022). The Torah seeks to protect biodiversity and perpetuate individual species that God created. "…God's purpose is carried through everything, even through a snake, a mosquito, a frog" (Bereishit Rabbah 10:7).

Judaism provides more than a "mind set" for sustainability, community well-being, and caring for our air, soil, and water. It instills moral responsibility providing rules for connecting people, community, and nature (Case & Chavez, 2017; Case, 2022). As stated in Ecclesiastes 1.4, "A generation goes and a generation comes, but the earth endures forever." The protection of the environment is sacred in Judaism. God appoints individuals (and institutions) as trustees to manage the environment and treat it with kindness, in co-stewardship of the world God created (Dorff & Ruttenberg, 2010; Sandelands & Hoffman, 2008). Humans alone can change and endanger the natural world through oil spills, toxic byproducts, and pollution. The biblical condition is that humans may use nature only in ways to enhance it.

Extended to the earth, the model reflects obligations of humanity toward the earth and its inhabitants as an obligation to God. Adam's right is also a responsibility. He is not given an unconditional mandate. Since the world belongs to God, man only has the right to use it according to God's desire. This includes a responsibility to develop the world for the better, to become partners with God's creation to guard, protect, and nurture the world. Humans are responsible to ensure the perpetuation and thriving of other species. Biological diversity and human distinctiveness are not mutually exclusive. The notion of an obligation to respond to others needs is the core of the covenantal foundation of Judaism (Tirosh-Samuelson, 2001, p 118) connecting the moral quality of human life to stewardship linked to injustice.

Heirs to this legacy and partnership are committed to alleviating environmental degradation of natural resource overuse; destruction of air, land, and water; wildlife protection; as well as suffering caused to human health and livelihood. Tirosh-Samuelson (2001) states that injustices arise from human greed and failure of people to protect the original order of creation. From a Jewish perspective, the just allocation of resources is a religious issue of the highest order (p. 19). Environmental justice is needed for future generations. "Choose life for you and your progeny" (Deuteronomy 30:19). By treating ecology in consonance with the deepest Jewish values, the well-being of God's created world is preserved for generations to come.

9.2.6 Climate Change: Global Warming

The Hebrew Bible includes rituals around seasons and harvests. In Deuteronomy 11:13–17, God promises rain as a gift to both the land and those who live on it if the Israelites heed God's commandments. God will "grant the rain for your land in season…I will provide grass in the fields for your cattle, and thus you will eat your fill." But if they turned away, God will "shut up the skies so that there will be no rain and the ground will not produce, and you will soon perish from the good land." Drought was the climate crisis of the ancient Mediterranean world. Rabbinic stories connect rain with earth and human survival. The ecotheologian Rabbi Arthur Waskow reminds contemporary Jews of these verses in Deuteronomy and the environmental consequences of their actions and inactions which lead to climate change. He points out that the rains become acid; oceans flood; and with that, the destruction of the ozone layer, "The sky itself will become your enemy" (2000, p. 269).

Because of human activity, the earth is reaching critical thresholds beyond which its ability to support life as we know it is at risk (Laszlo et al., 2020, p. 5). According to the Intergovernmental Panel on Climate Change (IPCC, 2022), human-induced climate change from industrial activity releases unprecedented levels of greenhouse gases into the air, raising global temperatures and sea levels and causing widespread damage to nature and people, beyond climate variability. The most vulnerable people and systems are disproportionately affected. Many vulnerable countries have scarcely contributed to global carbon emissions yet are paying the price of choices made by wealthier nations. Madagascar has only contributed 0.01% of the world's annual carbon emissions in the last 80 years but is currently experiencing the world's first climate famine (IPCC, 2022).

Driven by intersecting socioeconomic development, unsustainable ocean and land use, inequity, marginalization, and historical and ongoing patterns like colonialism and governance, somewhere between 3.3 and 3.6 billion people are highly susceptible to climate change as are a high proportion of species. Coral reef bleaching, ocean acidification, and deforestation lead to mass species extinction. Melting Arctic ice contributes to rising sea levels which will flood many coastal cities. Human and ecosystem vulnerability are interdependent, with a direct linear relationship between human-induced carbon dioxide accumulation and global warming (IPCC, 2018, 2022). The average temperature increased almost one degree Celsius in the last 100 years with some areas of the globe experiencing warming of more than 1.5 °C above pre-industrial levels leading to more severe weather patterns (IPCC, 2018). Things will worsen unless we remedy our infrastructure. In the sections that follow, I include Jewish ideas and practices that could help improve the world.

9.3 Improving the World

9.3.1 "Just Sustainability"

In Judaism, few values are as central as the obligation to do what is right and just (*tzedakah*), acts of kindness, mutual respect among groups, and the innate dignity of every human being (*gemilut hasadim*), and *tikkun olam*, healing and transforming the world (Case, 2014). The concept of *tzedakah* means righteousness, justice, and equity. It is a value which tries to correct the imbalances which humans create in society and in the natural world. Doing justice involves seeking a just world where you work to benefit others, treat them with compassion, and respect the dignity of all sentient beings, using your power for long-term good of people (Case, 2022). *Tikkun olam* combines both the opportunity to repair or protect the world and social justice applied to the natural world. This phrase originated in classical rabbinic literature, the *Mishnah*, in the third century, and the *Lurianic Kabbalah*, texts of Jewish mysticism, written in the mid-sixteenth century (My Jewish Learning, 2019). The concept now applies to all acts of contemplative, positive behavior. In the 1950s, it became part of the vocabulary of social justice where Jews are obligated to work toward creating a more perfect world, through human action, both individually and collectively (Noparstak, n.d.).

In our modern world, globalization allows free movement of people, information, money, goods, and services, simultaneously creating disruptions in local cultures and environments. While millions have acquired great wealth, millions more have not benefited. Instead, globalization has had a negative impact on both the environment and human rights. Capitalism includes assumptions that the more we grow, the better off we are with economic growth being an end in itself. Infinite growth is impossible on a finite planet (Laszlo et al., 2020, p. 4). Material consumption, and its benchmark the gross domestic product, is thought to be a pathway to happiness. An assumption that is free markets work best when unfettered by government regulation. The control of natural resources and the means of production including people's labor are in the hands of those on top (Eisler, 2007, p. 31; 2017b), called economic domination or "robber baron" capitalism by Rianne Eisler (2007, p. 117, 122). This leads to rising income inequality across the globe and increased damage to our ecosystem. The latter occurs because nature is viewed as an economic resource (Smith, 1937). We need a worldwide economy that is sustainable and equitable in resource and wealth distribution (Eisler, 2017a; Laszlo et al., 2020).

The four Jewish attitudes toward humanity and nature previously mentioned, which contribute to environmental sustainability and global justice, distinguish Judaism's different approach from unrestrained capitalism. Judaism is a religion of both strong individualism with the Jewish people linked to the religion through biological descent and strong communities with a collectivist social orientation emphasizing communitarian values (Cohn & Hill, 2007). Rabbi Hillel, a Jewish sage who lived at the end of the century BC (Goldin, 2012), makes clear this dual focus of the importance of the individual and community in his famous saying, "If I am not for myself, who will be for me? [individual responsibility] And being only for myself, what am I? [collective responsibility] And if not now, when?" (*Pirkei Avot* 1:14).

Environmental justice is a strong Jewish value (Neril & Dee, 2020, 2021). One cannot only focus on environmental conservation, referred to in most literature as environmental sustainability (Agyeman & Evans, 2004). Ignoring social and economic inequality that leaves marginalized communities bearing the brunt of toxic exposure and industrial pollution gravely violates the Jewish principle of justice. A recent environmental study concluded that minorities in the United States are exposed to disproportionately higher levels of air pollution from every source (industry, agriculture, vehicles, construction, residential) due to where their communities are located. They are often adjacent to industrial zones with diesel emissions from highways further polluting them leading to 85,000 to 2,000,000 excess deaths a year, the largest environmental cause of human mortality (Patnaik et al., 2020; Tessum et al., 2021). Deuteronomy 16:18 states, "Justice, justice shall you pursue." The Torah has repeated concerns for the welfare of the orphan, the widow, and the stranger emphasizing the moral imperative of protecting the poor and the vulnerable: "When one loves righteousness and justice, the earth is full of loving-kindness of the Eternal" (Psalms 33:5). The Torah is filled with other examples of "just sustainability" recognizing and resisting the collusion of racism, poverty, and environmental harm around the globe. This addition of race, class, and gender for social justice is considered more important in the environmental justice movement than only concerns for the environment (Agyeman et al., 2016).

Concern for "just sustainability" connects both the environment and social justice needs of individuals and communities as equally important (Jacobs, 2009; Mohai et al., 2009). The goal of "just sustainability" was redefined "to ensure a better quality of life for all, now, and into the future, in a just and equitable manner, while living within the limits of supporting ecosystems" (Agyeman et al., 2003, p. 2; Agyeman et al., 2016). Scripture includes rules about the treatment of the marginal in society (Case, 2022). Poor nations are disproportionately impacted by climate change (IPCC, 2022; United Nations, 2022; US Global Leadership Coalition, [USGLC], 2021). The effects of extreme weather and climate-driven disasters from floods and hurricanes are creating greater instability in fragile and emerging countries and markets driving mass displacement and threatening progress combating hunger, fighting global poverty, and preventing conflict. Environmental degradation is the key driver of fragility (USGLC, 2021). By 2050, more than 143 million people will be driven from their homes by conflict over food and water insecurity. By 2070, almost 20% of the planet will be too hot for habitation (USGLC, 2021). Warmer temperatures could expose over one billion people to deadly infectious diseases like Zika, dengue, and malaria (Dervis, 2007; USGLC, 2021). Countries most responsible for climate change should be those most accountable for finding solutions to confront the security, economic, and humanitarian consequences of climate change.

Development choices must prioritize risk reduction, equity, and justice. The IPCC, 2022 report states:

> Climate resilient development is enabled when decision-making processes, finance and actions are integrated across governance levels, sectors and timeframes involving international cooperation with governments at all levels working with communities, civil society, educational bodies, scientific and other institutions, media, investors, and businesses; and by developing partnerships with traditionally marginalized groups, including women, youth, indigenous Peoples, local communities and ethnic minorities. (D2)

Rabbi Jill Jacob's call to "just sustainability" demands "individuals refrain from actions that may cause harm to their immediate neighbors or to the community as a whole" (2009, p. 185). The Talmud makes a person who digs a pit on public property responsible for ensuing damages. Jacobs argues today this would include toxic runoff from factories, mountaintop removal mining, and other forms of industrial pollution. Such harm magnifies inequality.

9.3.2 Resource Conservation

9.3.2.1 The Sabbath and Sabbatical Year

For Martin Buber (1878–1965), the idea that God is the ultimate owner of the land is the cornerstone of the Jewish social concept, the Sabbath, and the sabbatical years (Dobb, 2000, pp. 37–38), important for resource conservation. Consider the ecological implications of the traditional application of laws for the Jewish Sabbath (Shabbat). It occurs on the seventh day of stillness after 6 days of Creation, beginning Friday evening at sundown and continuing through Saturday night until we see three stars (Belser, 2022, p. 9). This mirrors the act of Creation (Genesis 2:2). On Shabbat, one day out of seven, we alter how we interface with our environment by limiting our use of resources and restricting all types of productive activity. This includes not using electricity, operating machinery, lighting fires, or plowing fields. The Sabbath is a day where people walk instead of drive, carry no money, refrain from commerce, and do not shop or cook. Many Jews do not follow the Orthodox perspective. For those who follow it, Shabbat is a critique of the rat race of production and consumption. It is an antidote to the consumptive and utilitarian attitude toward land and its inhabitants that prevails during the week (Bernstein, 2018). The day is for relaxation, contemplation, and inner transformation where people encounter each other as a "Thou" (Buber, 1923, 2000). It is a sacred pause, even for the animals which can't be worked in the field or hunted (Exodus 20:10, 23:12; Deuteronomy 5:14). Spiritual renewal is affected through this physical contraction where one day of the week is for "unproductive rest," being, not doing, providing a sense of intimacy and spiritual connection with everything. The valuing of community and spirituality, which is reflected with the Jewish Shabbat, involves living a sustainable life where nature has a day of rest from human interference. This is true in how most Jewish holidays are observed.

The connection between land management, ritual, and social justice is evident in the laws regulating the Sabbatical year, a Shabbat for the land (*Shmita*) involving the resting or sustaining of the earth, analogous to the weekly Sabbath (Tirosh-Samuelson, 2001). This is another way Jews try to heal the world (Exodus 23:11; Leviticus 25:2–5; Deuteronomy 15:1–4). The literal meaning in Hebrew is "release." According to the Bible, every 7 years, the land should lie fallow (Leviticus 25:2–7). Before crop rotation and chemical enhancement of the soil, letting the land remain fallow enabled it to regain its fertility, allowing the earth to replenish itself with better harvests produced in the future. Just like a weekly day of rest, the earth also needs a Sabbath, so God instituted a sabbatical year. "Six years you shall sow your land and gather its produce; but in the seventh year you shall let your land rest and lie fallow. Let the needy among your people eat of it and what they leave, let the wild beasts eat" (Exodus 23:11). This commandment unintentionally had an ecological dimension. Every seventh year is committed to working on the ecology of the planet, being reminded of God's ownership of the land, with us but guests on earth. Sowing, harvesting, and gathering crops of any type is forbidden (Leviticus 25). The earth needs a rest from overuse of its resources, pollutants, and efforts to master and dominate it, so it can be rejuvenated. The Torah tells us that the earth will continue to produce food that we can gather and distribute, that this food has no owner, and that it is available to everyone (Lerner, 1994). Every seventh year, we are reminded we are not in control, just like earthquakes, floods, hurricanes, and wildfires remind us today. A sabbatical year stops ecological danger and allows new forms of responsible stewardship to emerge (Krantz, 2016). During that year, crops are not harvested by the landlord but are left for all to share including poor people and animals. *Shmita* combines ecology, social and economic justice, communal resilience, sustainable agriculture, and personal spirituality into an interconnected whole (Dobb, 2018).

Lerner (1994) suggests that this idea could be applied in two different ways. One involves a rotation, dividing the world into seven quadrants, with each taking a different sabbatical year. A second is we do this all together, reconnecting with the earth and one another. Jewish law retains its strong commitment to moral claims of community superseding property rights. The rabbis declared that the sin of Sodom was its inability to share its wealth with the stranger, the weak, and the poor (Lerner, 1994 p. 333).

9.3.2.2 Reforestation

Trees and forests have special reverence in the Bible. As mentioned earlier, the Torah is called a "tree of life," connecting trees to the highest Jewish values. A Midrash singles out trees in the Garden of Eden rather than the Garden itself to represent the natural world God created and the imperative not to destroy it (Ecclesiastes 7:1; Genesis 7:21–23). This example of *tikkun olam,* perfecting or repairing the world, reflects modern Jewish social justice theology. It is usually expressed as an activity to be done by humans in partnership with God. There is no one else to repair our world but us.

Another Midrash stresses this when the Jewish people enter the land of Israel. "It is said, 'Follow the Lord, your God (Deuteronomy 13:5). When he created the world, His first action was to plant trees, … 'And God planted a garden [of trees] in Eden (Genesis 2:8). So you, too, [follow His example], planting trees should be your first involvement'" (Vayikra Rabbah 25:3). Judaism even has a holiday, Tu BiShvat, where Jews celebrate the birthday of trees and rededicate themselves to the sacred values of environmental responsibility and stewardship.

A main concern of environmentalists today is the preservation of our natural resources like forests and arable land. The Torah addresses these concerns broadly stating actions that are destructive and wasteful are prohibited (NLE Olami Morasha, n.d., p. 15). The idea that the environment belongs to God includes this ethic of environmental responsibility, for instance, the Torah's demand in Parashat Shoftim: even during wars, fruit trees must be preserved by the opposing army when they take a city (Deuteronomy 20:19–20). From this biblical commandment, rabbis developed a protective legal umbrella encompassing ecological concerns for all types of needless destruction, forbidding unnecessarily wasting resources or destroying property (Wolff, 2009). Trees are compared to human beings. Destroying trees destroys human life, because it may destroy lives depending on them (Wolff & Neril, 2012). This interdependence is essential for ecological balance.

A *tikkun olam* effort initiated the Jewish National Fund for planting trees and forests in Palestine beginning in 1903, which continues today in Israel. Over the past century, the JNF has planted over 240 million trees, built over 180 dams and reservoirs, developed over 2500 acres of land, and created more than 1000 parks throughout Israel. Israel is the only nation in the world at the end of the twentieth century with more trees than it had at the beginning (Kornfeld, 2021, p. 1048). The trees planted were specifically selected to fit in the desert climate of Israel. They symbolize a healthy and sustainable environment, necessary for human life. Trees absorb more carbon than they release, slowing down global warming (Friedel, 2017). When trees are lost, like the 1000–2000-year-old sequoias in the 2020 and

2021 California wildfires, they are irreplaceable. What is required is a full investment in nature-based solutions for responsible reforestation.

9.3.2.3 Animal Welfare and Species Preservation

The Jewish attitude toward animals has always been governed by the command to confer dignity on all things God created considering that they, too, are God's creatures and intrinsically sacred (Kornfeld, 2021, p. 1030). The Midrash states that both Moses and King David were chosen by God to be leaders of Israel because of the compassion they had previously demonstrated toward their flocks (Shemos Rabbah 2:2). All living creatures are intrinsically sacred. In environmental terms, every species has inherent value beyond its instrumental use to human beings. The Torah demonstrates God's mercy to animals and the importance of not causing unnecessary pain with an entire code of laws (*tza'ar ba'alei hayim*) requiring the prevention of suffering of living creatures. They were to be treated with kindness and compassion (Regenstein, 2008). They were to be fed adequately (Deuteronomy 25:4) and treated properly (Exodus 23:4). In reading the story of the origin of mankind, Adam and Eve were not allowed to eat meat. It was not until Noah had saved the animal world from the Great Flood were he and his descendants allowed to kill animals for food (NLE Olami Morasha, n.d., p. 19).

Judaism includes rules legislating how animals used for food are raised and slaughtered requiring the elimination of unnecessary suffering (Regenstein, 2008; Rayner, 2003). Animals do not have rights, but we have duties toward them. One of those duties is not causing them pain or emotional distress. As previously discussed in the creation story, Genesis 1 gives us the mandate to "subdue" and "rule" creation, including animals. But Genesis 2 gives us the responsibility to "serve" and "guard." Since animals have feelings, we must respect them if we honor our role as God's partners in creation.

Unfortunately, 99% of production of chickens, turkeys, cows, and pigs are raised on factory farms in the United States as are 66% of the world's production of animals (Edgson, 2022). The meat industry has exponentially increased production and profits in a largely unregulated industry (Johnson, 2020). Factory-farmed animals are mostly confined to cramped spaces, where they are barely able to turn around or spread their wings and artificial environments where they are denied fresh air, sunlight, and exercise (Clark, 2022; Sears, 2014). More than 130 billion animals are forced to spend their lives in factory farms around the world enduring horrific, frightening, and unnaturally shortened lives (Clark, 2022).

Jewish law requires compassion to animals, disapproving of insensitivity to their natural needs and instincts. The usual way calves are raised for veal with newborn calves forcibly taken from their mothers, placed in small crates where they can barely move, and deprived of nourishment their entire brief life to make their flesh tender and pale (Regenstein, 2008, p. 9) is forbidden since it causes pain to the animals (Sears, 2014). At slaughter time, massive suffering is imposed on billions of animals every year as they are crammed into trucks with journeys of often 24 hours or more in extreme heat or cold, denied water, food, rest, or any relief. Jewish law considers this inhumane, prohibiting such practices (Regenstein, 2008; Sears, 2014).

A few factory farms are an exception, using humane methods of slaughter developed by animal scientist Dr. Temple Grandin who pioneered efforts to create standards that improve animal welfare, although such standards are not legally required (Grandin, 2020). Similarly, Kosher slaughter of animals seeks to minimize their pain. It originated in the Bible and has been observed by Jews for over 3000 years with most of the basic laws of kashrut derived from the Torah books of Leviticus and Deuteronomy (Neril & Dee, 2021). More sustainable kosher meat enterprises have emerged in the last 10 years with KOL Food and Grow and Behold Foods offering non-caged, grass-fed, antibiotic-free, glatt kosher meat. Their animals are raised in open pastures on small, family farms and then slaughtered under supervision of the Orthodox Union of Star K, with a mission to uphold the Jewish values of no cruelty to animals (Sears, 2014).

Jewish writings urge kindness toward all animals, stating that one who harms animals harms his own soul (Kornfeld, 2021, p. 1030). In the chicken industry, male chicks are considered useless so are killed immediately after hatching, sometimes stuffing them in plastic bags to suffocate them or grinding them up alive (Edgson, 2022). The average egg-laying factory farm has 800,000 birds confined together. These laying hens have their beaks removed without anesthesia to keep them from pecking one another to death in their tight quarters out of stress or anxiety (Clark, 2022; Edgson, 2022). These practices violate Jewish teaching since animals are treated like unfeeling commodities. If Jewish standards were followed, this would eliminate some of the suffering of animals.

Factory farming is also linked to the destruction of trees. It is estimated that 14,400 acres of rainforest are cut down daily to make space for cattle farms, destroying one acre every 6 seconds with environmental statistics showing 90% of the Amazon rainforest that have been cut down in the last 50 years are used in the meat industrial livestock process. This deforestation is connected to the extreme forest fires that lasted for months in 2019 and destroyed large parts of the rainforest (Edgson, 2022). If we were to treat the natural environment as a "Thou" rather than an "It," as Martin Buber suggested, we might be able to halt, or at least slow down, the degradation of our natural surroundings (Tirosh-Samuelson, 2001, p. 116).

Jewish law has no specific commandments protecting endangered species because people have the responsibility for all animals. Our ancestors would not have expected the biodiversity loss of the modern world (Troster, 2011).

9.3.3 *Urban Planning: Communal Responsibilities*

Since Jewish legal tradition is concerned with every aspect of society, many issues that we call urban planning are discussed. Cities were square, 1000 cubits (1500 feet) on each side. Greenbelts surrounded them, extending 500 cubits (750 feet) on each side with another square extending 500 cubits for fields and orchards. This was a small city of about seven blocks and a few hundred people. The greenbelt of 64 acres and 90 acres of fields could support this size city (Bentley, 2003, p. 47).

In urban communities concern over industrial pollution was expressed in *halacha*, assuming people who are given control over resources by God have discretion in how to manage them. Talmudic rabbis understood the interconnection of everything on earth, particularly the harm pollution causes. The Mishnah states: "Bits of broken glass should not be scattered on public land where they may cause injury" (Bava Kama 30a). There are also sources mandating proper disposal of waste and toxic products from industrial production that must be kept away from human habitation (Deuteronomy 23:13–15, Mishna Baba Batra 2:9). In Jewish tradition, public good overrides individual desires.

Pollution is never just local because it impacts places and people far away. *Pikuach nefesh* demands we consider the impact of our use of chemicals and other materials not only in the short term but also for the long term (Troster, 2011). Talmudic law placed limits on various trades and businesses with some barred from residential areas where they would not pollute the air or water supplies of city residents to ensure the health of people who lived there. This was the beginning of "zoning ordinances" (Bentley, 2003, p. 48).

Industries like tanning and dyeing had to be located at least 50 yards out of town, on the eastern edge, because the east wind is mild and reduces the unpleasantness of odors produced by tanning hides. There was a responsibility to guard against polluting water through these processes. Ovens producing clay utensils had to be located outside of towns so their smoke wouldn't pollute the air and cause fires presenting a danger to the town. Iron workers and blacksmiths were restricted to certain neighborhoods because of the dirt and noise they caused. Hazardous materials and vibration- and heat-producing machinery were limited to keep neighborhood communities safe. Even noise pollution was considered. Cemeteries also had to be at least 50 yards outside towns because of their potential to contaminate the water supply (Bentley, 2003, p. 48).

During the Hellenistic period, more Jews became urbanites, and laws from the Talmud affected them as well. Human and animal waste and garbage had to be carried away from the city so it would not be dispersed by wind or water. There were rules for facilities and private hygiene (Bentley, 2003). There were also other regulations in the Talmud for Jerusalem, the Holy City, that prohibited garbage heaps attracting rodents and lime and pottery kilns blackening buildings affecting residents' health (Zemer, 2003).

Jewish tradition offers insight into complex issues of modern times like zoning laws, minimum allotment for green space, regulations about garbage disposal and sewage sanitation, water pollution, and toxic waste. Fairness is a basic principle of Jewish law. Activities by governments, businesses, households, and other property owners may not cause harm to others. Based on the previously mentioned principle of *bal tashchit*, unnecessarily destroying resources of nature is not permitted. Much urban blight is caused by an economy driven by waste – too much packaging, single-use disposable products, private cars, large distances between homes and workplaces are issues urban design can address. Not all growth is healthy.

Establishing communities that care for its members and for the natural world is part of the Jewish tradition of sustainable development.

In an Accenture survey of Fortune 1000 CEOs, exploitive relationships that cannot be sustained have grown between the economy and the natural and social environment with globalization of industrial production (UN Global Compact, 2019). Sustainability as a strategy derives economic, ecological, and social value achieving both economic and environmental outcomes allowing companies to become agents of world benefit where businesses, the economy, and society thrive (Case & Chavez, 2017; Laszlo et al., 2020; Laszlo & Zhexembayeva, 2011). Sustainability recognizes global responsibility to others now and in the future, as does Jewish law.

9.3.3.1 Environmental Health

The principle of *pikuach nefesh*, saving human lives above all else, is one of Judaism's greatest moral obligations. We are taught, "You shall not stand idly by the blood of your neighbor" (Leviticus 19:16), and to "choose life, that you and your descendants may live" (Deuteronomy 30:20). Jewish values command preserving the earth and all forms of life for us now and for future generations. Divine concern is not measured in quarterly earning reports or electoral cycles. Jewish values involve long-term thinking that needs to be brought into the ecological discussion. It is our obligation to defend Creation and educate ourselves about the dangers of environmental health risks, working to prevent them for all humanity. In Judaism, dominion gives way to enlightened stewardship. Many factors affect global warming and climate change. Factory farming and environmental health are closely related.

Waste and Wastefulness Western society ritualizes consumption, whereas Jewish practice ritualizes reuse. There are consequences to society accustomed to disposability of things purchased like mountains of trash and oceans of plastic disrupting ecosystems (O'Neill, 2021). The Torah and Halakah prohibits wasteful consumption, reminding people that if they destroy the world, there will be no one left to repair what was (Ecclesiastes Rabba 7:13). As mentioned earlier, the Torah warns that even in times of war, trees must not be destroyed in a besieged city, with the prohibition going further, forbidding destruction of any object that could potentially benefit people. Even small amounts of food or fuel were to be used (Lerner, 1994, p. 328). The Jewish view is to first reduce use, then reuse, and only recycle as a last resort (Neril, 2012; Yoreh, 2019). Modesty in consumption is a value Jews have held for centuries (Troster, 2011).

Rabbis in later times extended this concept to not using more than what is needed and to using resources responsibly. If destroying trees for profit, do not destroy more than needed. If clear-cutting a forest interferes with the soil's ability to contain erosion or destroys an animal's habitat, the rabbis recommend relocating operations or altering techniques. Torah tradition does not recognize the right to unlimited consumption that in our consumer economy is focused on profit without regard to what we really need or what we are doing to our planet's resources.

A modern example of *bal tashchit* is discarding edible food. According to the Food and Agriculture Organization of the United Nations, "Roughly one-third of the food produced in the world for human consumption every year – approximately 1.3 billion tons- gets lost or wasted" (Gustavsson et al., 2011, p. v). In the United States, less than 3% of this waste is recovered or recycled. While some may be inedible, much is discarded by diners or are leftovers thrown away from refrigerators. Such food disposal costs approximately one billion dollars annually in the United States (Neril, 2012). Wasted food ends up in garbage dumps decomposing into methane, a destructive greenhouse gas. A study by Cuellar and Webber (2010) examined the energy costs for this wasted food including agriculture, transportation, processing, food sales, storage, and preparation. Costs represent about 2% of annual energy consumption in the United States. By wasting food, resources are squandered, reducing its availability to the world's poor. The Torah's call to not waste has ecological, social, and financial benefits (Neril, 2012).

The rabbis also created rules regarding air and water pollution (Mishna, *Bava Batra* 2:9; Talmud, *Yevamot* 44a). Maimonides (1190, 1974) points out in the *Guide of the Perplexed* that overexploited land erodes, losing its fertility. People are commanded to conserve the soil. Short-term gain should not be pursued at the cost of long-term desolation, reminding us we are responsible for sustaining the earth with a duty to use the world's resources judiciously and justly, a Jewish concept (Smith, 1991).

Clean Water Water has a respected place in Jewish tradition with an ethic for its protection. It plays a role in almost every major bible story. Isaac's wife was chosen at a well; (Genesis 24:1–14); Moses was saved as a baby from the bulrushes of the Nile River by the daughter of the pharaoh (Exodus 2:1–10); the Israelites were freed from slavery in Egypt when the Red Sea parted (Exodus 14:21); and Miriam gave water to the Jewish people in the desert (Numbers:21). It also plays a role in ritual use like full-body immersion in a ritual bath, ritual handwashing before eating, and a water ritual at Rosh Hashanah, the Jewish New Year, called *Tashlich*, where you symbolically throw your sins into water, a renewal ritual. Since water is an important resource for peoples' survival, Jewish laws against water pollution threatening community health and economic well-being ensure clean water is available for everyone (Troster, 2018).

Water covers 70% of the Earth's surface, yet freshwater comprises less than 1% of this. Currently, over 800 million people lack access to clean drinking water, a condition that will only worsen (Troster, 2018) with the world's supply of freshwater being exhausted with industrial and population growth. As populations grow, so does water usage, sewage-contaminated water, polluted agricultural runoff, and contaminated water from industrial, commercial, and residential uses. Waste from raising cattle, poultry, and hogs has polluted 145,000 miles of rivers and streams in the United States as well as a million acres of lakes, reservoirs, and ponds (Edgson, 2022). According to the EPA, one pound of meat produced consumes 2400 gallons of water leading to water wastage, whereas one pound of wheat uses only 25 gallons (Edgson, 2022).

Coca-Cola in rural villages in India used village ground water to produce its beverages, in effect drying community wells causing great harm to people's health and economic well-being. The bottling process then produced waste and was discarded back into the community water supply, leaving remaining freshwater sources too polluted for consumer use in agriculture (Poonamalee & Howard, 2009). This is not a morally responsible practice. Jewish law would not allow this since protecting the ecosystem is important in Judaism.

9.3.4 *Tikkun Olam*

In modern Jewish circles, the phrase *tikkun olam* has become synonymous with social action providing its framework for social justice work. It involves righteous behavior, justice, and fairness. In Judaism, God invites us to be his "partners in the work of creation" (Pirke Avot). The protection and fixing of the world are in our hands. Social responsibility is a collective responsibility over time. The religious imperative is to create, not to destroy, done one day at a time, one person at a time, one act at a time (Sacks, 2005) with justice and compassion (Hertzberg, 1986). There are many Jewish communal organizations whose guiding mission is to help alleviate poverty, hunger, and disease throughout the world, regardless of race, religion, or nationality. They improve people's chances of survival, providing economic independence and enhancing human dignity (Dobb, 2018).

9.4 Judaism's Contributions to Sustainability Dialogue

9.4.1 *Saving the Planet*

Judaism offers key contributions to the current sustainability dialogue. The examples provided in this chapter help establish that environmental efforts are an authentic, organic, outgrowth of Jewish teaching and tradition, showing how Jewish values can align with and help drive social change. In 1836, Rabbi Samson Raphael Hirsh stated, "They ask you to regard all living things as God's property. Destroy none; waste nothing; employ things wisely…Look upon all creatures as servants in the household of creation" (Hirsch, 1836, 1995). Similarly, Rabbi Jonathan Sacks interprets the oral tradition in Judaism of "do not destroy" as expanding the horizons of corporate moral responsibility to unborn generations (Sacks, 2016). Judaism proscribes ways for civil societies to promote living in harmony with nature and protecting it for future generations. Making sure "the earth endures forever" means implementing policies that encourage conservation and a greater reliance on renewable energy.

We need to develop a plan to use the world's resources and distribute the world's wealth. Currently wealthy countries buy resources from poorer countries and shape them into consumer products fitting within our capitalistic notion of limitless consumption. This increases planetary pollution, escalating destruction of the ecosystem. There is a disproportionate asymmetry of the exporting of recyclable waste from developed countries to developing countries because it is cheaper to export rather than develop local recycling infrastructures, reducing landfilling, and is lucrative for importers (Martínez et al., 2022; Tieso, 2022). These developing countries are not equipped to handle the mass import of waste in an environmentally sound manner, as well as lack resources to mitigate its health impacts (Martínez et al., 2022; Sonak et al., 2008). A plan needs to address what the world wants to be produced and what resources need to be preserved for future generations. It also needs to consider restrictions on production to ensure the environment is protected including what needs to be done with toxic waste. Saving the planet needs to be our priority (Dobb, 2018). Shabbat and the sabbatical year are profound critiques of production and consumption and a reminder that we are at our best when we enjoy rather than make – "being" not "doing." They serve important practical contributions to this effort.

Any strategy to save the planet must also address social justice issues with developing countries participating. But for them to do that, the imbalances of wealth between the developed and the underdeveloped world need to be remedied (Dhar & Case, 2023). A Jewish environmental outlook evinces a concern for every person with a special focus on those least advantaged, applying an environmental justice mindset. The future of humanity depends on how wisely resources are used in the world. Redistribution of the world's wealth is required so that poverty will not continue to lead to ecological destruction in the Third World. We must also recognize that advanced industrial societies are responsible for the misuse of most of the world's resources and its pollution. Shabbat and the sabbatical year in Jewish practice introduces reconnecting to the earth's rhythms instead of challenging ways we subdue and shape the world to our immediate needs (Dobb, 2018).

It isn't enough to change individual behavior. There are economic and political realities that shape choices throughout the world. One of the heaviest polluters in the world is the automobile. Both the oil and auto industries worldwide have blocked mass transportation systems that are more ecologically friendly lobbying against climate change legislation and regulation to tackle global warming (McCarthy, 2019; Parafiniuk & Smith, 2019; Stephen et al., 2019). They want people to be dependent on cars (Bronsdon, 2021). These industries use their resources to elect legislators sympathetic to their industry needs. For instance, in 2018, Koch Enterprises funded Americans for Prosperity, a political group, to encourage voters to oppose proposed public transit plans across America (Tabuchi, 2018). System change is necessary (Jacobs, 2009; Rose et al., 2008). With larger transformation necessary to save the planet, the Jewish perspective that God rules the universe enables transformation when people feel powerless to make a difference (Tirosh-Samuelson, 2019, 2022). Judaism brings a communitarian spirit and long-term perspective into our contemporary discourse (Case, 2022). It focuses on providing from one generation to the next, "choose life, that you and your descendants may live" (Deuteronomy 30:19).

Another massive contributor to our inability to preserve our world is our modern agri-business for food consumption. The way we currently do this contributes to global warming, soil erosion, air and water pollution, overuse of chemical fertilizers and pesticides, destruction of tropical forests and habitats, and other environmental damage (Regenstein, 2008, p. 12–13).

How do we transform the market economy that challenges self-interest to take stewardship seriously? Many multinational organizations have adopted sustainable strategies to lessen the impact of their businesses on the planet. Some focus on sustainability initiatives, while others have sustainability goals for health, poverty, education, and equal access to resources (Bleich, 2020). Multinational companies doing this work include Accenture, Adidas, Apple, Banco de Brasil, Bosch, Estee Lauder, GlaxoSmithKline, Google, Ikea, Intel, Lego, and Patagonia (Bleich, 2020; Port, 2021; CleanRiver, 2020). Some track environmental impacts of water and land use, energy efficiency, chemical use, greenhouse gasses, and labor policies. Others improve sustainability around food yields and eliminate antibiotics in animal production. They reduce carbon emissions, invest in biofuels, run on 100% renewable green energy, send no waste to landfills recycling it into energy, use sustainable materials, erase plastic from their supply chain, provide green loans for environmentally focused projects, and donate earnings to environmental organizations (Bleich, 2020; Port, 2021; CleanRiver, 2020). Apple works with Conservation International replanting 27,000 mangrove trees along the Columbia's coastline to "scrub" carbon from the atmosphere and protect it from erosion; Google uses herds of goats to trim their grass as well as greened their supply chain; and Bosch has dedicated 50% of its research and development budget creating and supporting technology protecting the environment (Bleich, 2020). These mainstream companies realized they can thrive while reducing environmental harm. More companies need to be concerned about dangers to the earth's ecology by ceaseless pursuit of economic gain. All can be stewards of the planet with a responsibility to take care of and guard nature. They do well from doing good, being "partners in the work of creation."

9.4.2 The Jewish Ecological Movement

The Jewish ecological movement was started by Rabbi Ellen Bernstein 34 years ago, through an organization she founded, *Shomrei Adamah*-Keepers of the Earth (Bolton-Fasman, 2020). Consciously caring about the world is an essential part of Jewish identity (Bernstein, 2018). The Jewish passion for justice, with its ethics of responsibility, was extended to the physical environment endeavoring to protect people and other species from environmental degradation (Tirosh-Samuelson, 2001). This movement educated Jews and inspired them to lead environmentally sound lives based on Jewish values expressed in the sacred sources of Judaism. Although Judaism is consistent with many contemporary environmental doctrines, its teachings are not identical to them. Certain assumptions, widely taken for granted by secular environmentalists, conflict with Jewish tradition. For instance, Jewish environmental ethics is not based on a simplistic version of pantheism acknowledging the entire universe in its idea of God. Biocentrism, from a Jewish perspective, is a form of paganism that results in an idolatrous worship of nature (Schwartz, 1995). An environmental philosophy that just reveres what is, while ignoring what should be, is not viably Jewish. In Jewish teaching, love and respect of people goes hand in hand with human responsibility toward other species created by God. In Judaism, the entire world is interconnected. There is a need to emphasize empathy, build relationships and coalitions, and develop mechanisms fostering interdependence (Dobb, 2018).

The preservation and protection of nature is important, but not the most important societal value. It sees humans as more than part of nature, but as privileged with moral claims. It allows nature to be used and enjoyed but also insists it be protected. The key contribution of ancient Jewish texts to contemporary environmental discourse is the intricate system of expected behavior around the concept of balance between human needs and nature as a source of life, with both having purpose. Ecological and sustainable needs in Judaism also focus on structures of oppression, exploitation, and environmental degradation. Humans can use nature for their own benefit. But it does not exist only for their benefit. As God's creation, it has value independent of human needs. Although there are many differences in our contemporary environmental challenges to those faced during the time of development of Jewish Scripture which focused on total engagement with the world, there are many "green" elements in them, with the central principle of "do not destroy." This can inspire sound environmental policies for actions needed for intervening effectively in the world, becoming part of the solution to the contemporary environmental crisis (Tirosh-Samuelson, 2001).

Liberal Jews (who may be Reform, Reconstructionist, or secular) adopt religious law to their modern lives. They have more progressive attitudes toward the environment, willingly cooperating with all groups working to improve the state of the world. On the other hand, those who are Orthodox live by the Torah and rabbinic law (Halakha), using it to guide their actions. They are more likely to believe that caring for their fellow humans takes priority over the environment. This is because they believe God gave humans dominion over the world with a responsibility to bring it under their control and be responsible for its maintenance.

Within that sacred book is groundwork for social reform: humans have the ability and therefore responsibility to stop injustices in the world. They join with environmental groups and actions that can be supported from the Torah focusing on social justice and social harmony, performing *mitzvot* which deals with relationships of people to God and each other which is more important to them than how they treat nature. Linking sustainability dialogue to a fight against inequality in the world allows them to do the work required in the Torah.

Judaism contains both "green" and "non-green" elements, but it is not a "green" religion. The biblical imperative requires finding a balance between transformation of nature and its preservation. This responsibility is on us. God doesn't "do." We human beings "do." It is our responsibility to act with God to bring Godliness into the world, to be the good that overtakes evil. We have the power to be God's partner in this endeavor.

9.4.3 Implications

With our planet facing an ecological crisis, the teachings of Judaism provide an approach to address environmental problems threatening our future in an interdependent manner, valuing the environment, the world's people, God, and economic progress. Judaism is rich with environmental stewardship mitzvah like *bal tashchit* where you do not destroy and limit resource use. Duties require the rescue of innocent people from injury or death (*hatzala*, the value of rescue). There is also the integration of God, where you are commanded to love God as you do your neighbors, living in ways God would find acceptable. Judaism elevates mundane aspects of life, making everything sacred. The focus is on what needs tending or fixing (Weintraub, 2022).

What do you do when the rewards of environmental protection, species preservation, and human health come at the expense of global trade, job creation, or economic opportunity? The Jewish religious perspective is particularly suited to this kind of tension, contributing to contemporary efforts addressing the ethical relationship between people and the natural world in which we live.

It is critical for environmentalists and people of faith traditions to work together toward the preservation of the world including the health of the land and a greater level of social and economic equity, both within and between nations. The Jewish religious establishment offers a powerful infrastructure for long-term objectives to promote its environmental message of "just sustainability." Along with other faith communities, masses of people can be organized to do the same. Yet the religious and secular worlds often do not understand each other. In this chapter, I developed a case for the use of Jewish sacred texts in our ecological discourse to establish a common bond between the followers of biblical traditions, conservationists, environmentalists, and others who care about a more sustainable world.

Judaism's historical commitment to environmental issues has a concern for human welfare. Religious sources present multiple perspectives, in tension and balance, that examine both sides of the tension, with an awareness of the nuances of cost-benefit analysis (Mikva, 2003, pp. 35–36). From a Jewish perspective, health concerns override most issues with environmental debates often involving questions of human health like pesticides, industrial pollution, ozone depletion, global warming, and even biodiversity (Mikva, 2003, p. 42).

Itzchak Kornfeld (2021) points out that the Jewish canon offers prescriptions for what ails the earth: a *tikkun olam* ("repairing the world" or "making the world better") that is thousands of years old. There is guidance offered about how to be "guardians" of the earth, rather than trying to control and dominate it (pp. 1050–51). Whether one believes in God or not, there is a guidebook, the Torah, providing direction for safeguarding our planet and its people. Halakhic discussion goes beyond narrow legal questions, integrating larger ethical considerations and individual conscience. What is missing is a "willingness to do so" by human beings. Judaism focuses on a collaborative mutuality of the covenant with God and with one's fellow man. Each member of society must accept responsibility for communal burdens while also benefiting from a functional infrastructure. Environmental burdens and benefits need to be distributed equitably without regard for the power and wealth of affected communities. Jewish law doesn't tell us how to govern states or global corporations but insists on individual responsibility for community action and responsibility (city, state, federal government) for the sustainability of life within communities and for the health and welfare of its members (Case, 2022).

Joining the macro- and micro-focuses together could lead to developing energy-efficient, affordable housing, close to public transportation. It could lead to the creation of jobs paying living wages through enhanced recycling services, composting of organic waste, and each community (privileged or vulnerable) sharing an equitable burden for preventing and disposing of toxic waste. The Jewish texts drawn on balance concern for the welfare of humanity while also protecting the natural world. Judaism can make useful contributions to ecological discourse, inspiring ecological policies (Tirosh-Samuelson, 2019). The challenge of our times is bringing together conservation people with those who have environmental justice concerns. Judaism does that with its ideas of equality, shared responsibility among people, and preservation of the natural world for generations to come (Jacobs, 2009, pp. 191–192).

Together we can weave together a new story of people and land, ensuring a way of life that can sustain the land, its creatures, and us as a species. We are all in this together. Like Dr. King's "fierce urgency of now" (Martin Luther King, 1963), we must ask with Hillel: "If not now, when?" (Pirkei Avot 1:14).

References

Agyeman, J., & Evans, B. (2004). 'Just sustainability': The emerging discourse of environmental justice in Britain. *The Geographical Journal, 170*(2), 155–164. https://doi.org/10.1111/j.0016-7398.2004.00117x

Agyeman, J. Bullard, R., & Evans, B. (Eds.) (2003). Introduction: Joined-up thinking: Bringing together sustainability, environmental justice, and equity. *Just sustainabilities: Development in an unequal world* (pp. 1–16). MIT Press.

Agyeman, J., Schlosberg, D., Craven, L., & Matthews, C. (2016). Trends and directions in environmental justice: From inequity to everyday life, community, and just sustainabilities. *Annual Review of Environment and Resources, 41*, 321–340. https://doi.org/10.1146/annurev-environ-110615-090052

Andersen, L., Corazon, S. S., & Stigsdotter, U. K. (2021). Nature exposure and its effects on immune system functioning: A systematic review. *International Journal of Environmental Research and Public Health, 18*(4), 1416. https://doi.org/10.3390/ijerph18041416

Aycrigg, J. L., Mccarley, T. R., Belote, R. T., & Martinuzzi, S. (2022). Wilderness areas in a changing landscape: Changes in land use, land cover, and climate. *Ecological Applications, 32*(1), 2471. https://doi.org/10.1002/eap.2471

BBC. (n.d.). *Sustainability and unsustainability: Some facts.* BBC. https://www.bbc.co.uk/bitesize/guides/zr3c7ty/revision/1

Belser, J. W. (2022). A tree of life: Torah and interpretation in Jewish tradition. In H. Marlow & M. Harris (Eds.), *The Oxford handbook of the Bible and ecotheology*. Oxford University Press.

Bentley, P. J. (2003). Urban planning in Jewish environmental law. In W. Jacob & M. Zermer (Eds.), *The environment in Jewish Law: Essays and responsa* (pp. 45–55). Berghahn Books.

Bernstein, E. (2018). A biblical land ethic? A response to Aldo Leopold. In M. Brotton (Ed.), *Ecotheology in the humanities: An interdisciplinary approach to understanding the divine and nature*. Lexington Books. https://www.thepromiseoftheland.com/a-biblical-land-ethic

Berrett, D. (2011). Economists push for a broader range of viewpoints in their field. *The Chronicle of Higher Education, 13*, A11. https://www.chronicle.co…

Bleich, C. (2020). 12 examples of corporate sustainability initiatives to inspire your goals. *Edgepoint Learning*. https://www.edgepointlearning.com/blog/examples-of-corporate-sustainability-initiatives/

Bolton-Fasman, J. (2020, March 23). Greening the Passover Haggadah. *Jewish Boston*. https://www.jewishboston.com/read/greening-the-passover-haggadah/

Borrelli, P., Robinson, D. A., Fleischer, L. R., Lugato, E., Ballabio, C. A., Meusburger, K., Modugno, S., Schutt, B., Ferro, V., Bagarello, V., Van Oost, K., Montanarella, L., & Panagos, P. (2017). An assessment of the global impact of 21st century land use change on soil erosion. *Nature Communications, 8*, 1–13. https://doi.org/10.1038/s41467-017-02142-7

Brauner, R. (2011). *Thinking Jewish: The art of living in two civilizations*. Mirkov Publications.

Bronsdon, C. (2021). Electric cars won't solve climate change. *Planetizen*. https://www.planetizen.com/blogs/112490-electric-cars-wont-solve-climate-change

Buber, M. (2000). *I and Thou*. (R. G. Smith., Trans.). Scribner. (Original work published 1923).

Calil, J., Gutiérrez-Graudiņš, M., Munguía, S., & Chin, C. (2021). *Neglected: Environmental justice impacts of marine litter and plastic pollution*. United Nations Environment Programme. https://www.unep.org/resources/report/neglected-environmental-justice-impacts-marine-litter-and-plastic-pollution

Case, S. S. (2014). Religious identity salience, values, and career. In S. Case (Ed.), *USSO 289C: Ethics for the real world*. Case Western Reserve University.

Case, S. (2022). Morally responsible behavior in unprecedented times: Relevance of sacred texts of Abrahamic religions to workplace behavior. In M. Chi Vu, N. Burton, I. Chu, & N. Singh (Eds.), *Faith traditions and practices*, pp. 151–181. Palgrave Macmillan.

Case, S. S., & Chavez, E. (2017). Guiding lights for morally responsible sustainability in organizations: Revisiting sacred texts of Judaism, Christianity, and Islam. *Journal of Organizational Psychology, 17*(2), 35–49.

Case, S., & Smith, J. (2013). The genesis of integrity: Values and virtues illuminated in Judaism, Christianity, and Islam for workplace behavior. In W. Aman & A. Stachowicz-Stanusch (Eds.), *Integrity in organizations: Building the foundations for humanistic management* (pp. 307–344). Palgrave Macmillan.

Chapin, S., Carpenter, S. R., Kofinas, G. P., Folke, C., Abel, N., Clark, W. C., Olsson, P., Smith, D. M., Walker, B., Young, O. R., Berkes, F., Biggs, R., Grove, J. M., Naylor, R. L., Pinkerton, E., Steffen, W., & Swanson, F. J. (2010). Ecosystem stewardship: Sustainability strategies for a rapidly changing planet. *Trends in Ecology & Evolution, 25*(4), 241–249. https://doi.org/10.1016/j.tree.2009.10.008

Chen, J., & Mueller, V. (2018). Climate change is making soils saltier, forcing many farmers to find new livelihoods. *The Conversation*. https://theconversation.com/climate-change-is-making-soils-saltier-forcing-many-farmers-to-find-new-livelihood2022)s-106048

Clark, S. (2022, September 13). *Industrial agriculture is far worse than you think.* The Humane League. https://thehumaneleague.org/author/shad-clark

CleanRiver. (2020). *What companies use sustainable business practices? 8 sustainable companies across the globe*. https://cleanriver.com/8-sustainable-companies-across-the-globe/

Cohn, A. B., & Hill, P. C. (2007). Religion as culture: Religious individualism and collectivism among American Catholics, Jews, and Protestants. *Journal of Personality, 75*(4), 709–742. https://doi.org/10.1111/j.1467-1494.2007.00454.x

Cuellar, A. D., & Webber, M. E. (2010). Wasted food, wasted energy: The embedded energy in food waste in the United States. *Environmental Science & Technology, 44*(16), 6464–6469. https://doi.org/10.1021/es100310d

Dervis, K. (2007). Devastating for the world's poor: Climate change threatens the development gains already achieved. *Green Our World, XLIV*(2). UN Chronicle. https://www.un.org/un/chronicle/article/devastating-worlds-poor-climate-change-threatens-development-gains-already-achieved

Dhar, U., & Case, S. (2023). A postcolonial deconstruction approach towards promoting socially-conscious management in emerging economies. In L. Poonamallee, A. Howard, & S. Joy (Eds.), *Managing for social justice: Harnessing management theory and practice for collective well-being, pp.15-40.* Palgrave Macmillan.

Dobb, F. (2000). Four modern teachers: Hirsch, Kook, Gordon, and Buber. In A. Waskow (Ed.), *Torah of the earth: Exploring 4,000 years of ecology in Jewish thought* (Vol. 2, pp. 21–41). Jewish Lights Publishing.

Dobb, F. (2018). The Jewish basis for environmentalism. *Evolve: Groundbreaking Jewish Conversations.* https://evolve.reconstructingjudaism.org/the-jewish-basis-for-environmentalism/

Dorff, E. N., & Ruttenberg, D. (Eds.). (2010). *Jewish choices, Jewish voices: Social justice.* The Jewish Publication Society.

Edgson, J. (2022, August 15). *41 dreadful factory farming statistics (2022 Update).* Petpedia. https://petpedia.org/factory-farming-statistics…

Eisler, R. (2007). *The real wealth of nations.* Berrett-Koehler Publishers, Inc.

Eisler, R. (2017a). Roadmap to a caring economics: Beyond capitalism and socialism. *Interdisciplinary Journal of Partnership Studies, 4*(1), Article 3. https://doi.org/10.24926/ijps.v4i1.149

Eisler, R. (2017b). The real wealth of nations: From global warming to global partnership. *Interdisciplinary Journal of Partnership Studies, 4*(3), Article 13. https://pubs.lib.umn.edu/ijps

EPA. (2021). *EPA report shows disproportionate impacts of climate change on socially vulnerable populations in the United States.* United States Environmental Protection Agency. https://www.epa.gov/cira/social-vulnerabilit-report

Fougères, D., Jones, M., McElwee, P., Andrade, A., & Edwards, S. (2022). Transformative conservation of ecosystems. *Global Sustainability, 5*(E5). https://doi.org/10.1017/sus.2022.4

Friedel, M. (2017, July 18). Forests as carbon sinks. *American Forests.* https://www.americanforests.org/article/forests-as-carbon-sinks/

Friedman, M. (1962). *Capitalism and freedom.* University Of Chicago Press.

Goldin, J. (2012, January 15). Hillel. *Encyclopedia Britannica.* https://www.britannica.com/biography/Hillel

Grandin, T. (Ed.). (2020). *Improving animal welfare: A practical approach* (3rd ed.). CABI. https://styluspub.pressurewsarehouse.co…

Gustavsson, J., Cederberg, C., Sonesson, U., van Otterdijk, R., & Meybeck, A. (2011). *Global food losses and food waste – Extent, causes, and prevention.* Food and Agricultural Organization of the United Nations. https://www.fao.org/3/mb060e/mb060e01.pdf

Hertzberg, A. (1986). The Jewish declaration on nature. *The Assisi declarations: Messages on humanity and nature from Buddhism, Christianity, Hinduism, Islam, and Judaism.* World Wildlife Federation, 25th Anniversary, 29 September. https://www.silene.ong/en/documentation-centre/declarations/the-jewish-declaration-on-nature

Hirsch, R. S. R. (1836). *The nineteen letters: The world of Rabbi S. R. Hirsch.* (J. Elias, Trans. 1995). Feldheim Publishers.

Hoffman, A. J., & Sandelands, L. E. (2005). Getting right with nature: Anthropocentrism, ecocentrism, theocentrism. *Organization and Environment, 18*(2), 141–162.

IPCC. (2018). Summary for policymakers. In: *Global Warming of 1.5°C. An IPCC Special Report on the impacts of global warming of 1.5°C above pre-industrial levels and related global greenhouse gas emission pathways, in the context of strengthening the global response to the threat of climate change, sustainable development, and efforts to eradicate poverty.* Cambridge University Press, (pp. 3–24). https://doi.org/10.1017/9781009157940.001

IPCC. (2022). *Climate change 2022: Impacts, adaptation and vulnerability.* Contributions of Working Group II to the Sixth Assessment Report of the Intergovernmental Panel on Climate Change. Cambridge University Press. https://www.ipcc.ch.report

Islam, S. N., & Winkel, J. (2017). *Climate change and social inequality*. United Nations Department of Economic & Social Affairs. https://www.un.org/esa/desa/papers/2017/wp152_2017.pdf

Jackson, J. (2022, January 12). Repairing our world: Jewish environmentalism through text, tradition, and activism. *Folklife Magazine.* Smithsonian Center for Folklife and Cultural Heritage. https://folklife.si/edu.ma…

Jacobs, J. (2009). *There shall be no needy: Pursuing social justice through Jewish law and tradition*. Jewish Lights Publishing.

Johnson, A. (2020, December 1). *Factory farming: What it is and why it's a problem.* The Humane League. https://thehumaneleague.org?article?what-is-factory-farming…

King, M. L. (1963, August 28) *I have a dream speech*. Washington, DC. http://www.huffingtonpost.com/2012/01/16/i-have-a-dream-speech-text-martin-luther-king-jr_n_12207734.htm

Kornfeld, I. E. (2021). Let the earth teach you Torah: Sustainability in Jewish law. *Touro Law Review, 36*(4), 1003–1051.

Krantz, D. (2016). Shmita revolution: The reclamation and reinvention of the sabbatical year. *Religions, 7*(8), 100.

Krone, A. (2021, June 23). Jewish environmentalism. *Shalvi/Hyman Encyclopedia of Jewish Women.* Jewish Women's Archive. https://jwa.org/encyclopedia/article/environmentalism

Landrigan, P. J., Rauh, V. A., & Galvez, M. P. (2010). Environmental justice and the health of children. *Mount Sinai Journal of Medicine, 77*(2), 178–187. https://onlinelibrary.wiley….

Landrigan, P. J., Fuller, R., Acosta, N. J. R., Adeyi, O., Basu, N., Balde, A. B., Bertollini, R., Bose-O'Reilly, S., Boufford, L., et al. (2018). The Lancet Commission on pollution and health. *Lancet, 391*, 462–512. https://doi.org/10.1016/S0140-6736(17)32345-0

Laszlo, C., & Zhexembayeva, N. (2011). *Embedded sustainability: The next big competitive advantage*. Stanford University Press.

Laszlo, C., Cooperrider, D., & Fry, R. (2020). Global challenges as opportunity to transform business for good. *Sustainability, 12*(8053), 1–17. https://doi.org/10.3390/su1219853

Lerner, M. (1994). *Jewish renewal: A path to healing and transformation*. GP Putnam's Sons.

Lim, T. H. (2013). *The formation of the Jewish Canon*. Anchor Yale Bible Reference Library.

Lock, M. K. (2021, February 25). What does the Bible say on the preservation of the environment? *The Jerusalem Post*. https://www.jpost.com/judaism/what-does-the-bible-say-on-the-preservation-of-the-environment-660189

Maimonides, M. (1190). *The guide of the perplexed, Vol. 1.* (S. Pines, Trans. 1974). University of Chicago Press.

Martínez, J. H., Romero, S., Ramasco, J. J., et al. (2022). The world-wide waste web. *Nature Communications, 13*, 1615. https://doi.org/10.1038/s41467-022-28810-x

Maslin, M. (2004). *Global warming. A very short introduction*. Oxford University Press.

McCarthy, N. (2019). Oil and gas giants spend millions lobbying to block climate change policies. *Forbes*. https://www.forbes.com ›mccarthy

Mikva, R. S. (2003). When values collide: Economics, health, and the environment. In W. Jacob & M. Zermer (Eds.), *The environment in Jewish law: Essays and responsa* (pp. 34–44). Berghahn Books.

Mohai, P., Pellow, D. N., & Roberts, J. T. (2009). Environmental justice. *Annual Review of Environment and Resources, 34*, 405–430. https://doi.org/10.1146/annirev-environ-082508-094348

Munch, D. (2021). *2020 disaster estimations reach at least $3.6 billion in uncovered loss*. Market Intel. American Farm Bureau. https://www.fb.org/market-intel/2020

MyJewishLearning.com. (2019). *Tikkun olam: Repairing the world.* https://www.myjewishlearning.com/article/tikkun-olam-repairing-the-world/

Neril, Y. (2012, August 1). *Holy use: Relating to resources sustainably.* Hazon. https://hazon.org/holy-use-relating-to-resources-sustainably/

Neril, Y. (2014). *Countering destruction: Lessons from Noach*. Canfei Nesharim. http://canfeinesharim.org/countering-destruction-lessons-from-noach-longer-article/

Neril, Y., & Dee, L. (2020). *Eco Bible: Volume 1: An ecological commentary on Genesis and Exodus*. The Interfaith Center for Sustainable Development.

Neril, Y., & Dee, L. (2021). *Eco Bible: Volume 2: An ecological commentary on Leviticus, Numbers, and Deuteronomy*. The Interfaith Center for Sustainable Development.

NLE Olami Morasha. (n.d.). *Morasha syllabus, The Jewish View of Ecology and the Environment*. nleresources.com. https://www.nleresources.com/morashasyllaus/the-jewish-view-of-ecology-and-the-environment

Noparstak, J. (n.d.). Tikkun olam. *Learning to give*. https://www.learningtogive.org/resources/tikkun-olam

O'Neill, G. (2021). U.S. plastic pollution & waste export. *UNA-NCA Snapshots*. https://medium.com/una-nca-snapshots/u-s-plastic-pollution-waste-export-41111f195b50

Parafiniuk, A., & Smith, Z. A. (2019). Green gilded oil: How faux sustainability by US oil companies is undermining neo-sustainability. *Sustainability, 11*(14), 3760. https://doi.org/10.3390/su11143760

Patnaik, A., Son, J., Feng, A., & Ade, C. (2020). *Racial disparities and climate change*. Princeton University. https://psci.princeton.edu/tips/2020/8/15/racial-disparities-amd-climate-change

Poonamalee, L., & Howard, A. (2009). The dark side of water: Struggle for access and control: A case. In E. Raufflet & A. Mills (Eds.), *The dark side – Critical cases on the downside of business* (pp. 206–221). Greenleaf Publishing.

Port, S. (2021, December 8). *The Adidas sustainability story – Leading the change*. GamePlan A. https://www.gameplan-a.com/2021/12/the-adidas-sustainability-story-leading-the-change/

Rayner, J. D. (2003). Judaism and animal welfare. In W. Jacob & M. Zermer (Eds.), *The environment in Jewish law: Essays and responsa* (pp. 56–72). Berghahn Books.

Regenstein, L. (2008). *Commandments of compassion: Jewish teachings on protecting animals and nature*. The Humane Society of the United States. https://www.humanesociety.org/sites/default/files/archive/assets/pdfs/faith/commandmentsofcomp_3_08.pdf

Rose, N., Kaiser, J., Klein, M., & (Eds.). (2008). *Righteous indignation: A Jewish call for justice*. Jewish Lights.

Rousseau, J. (1762/1968). *The social contract*. (G. D. H. Cole, Trans. 1968). Penguin. Original published in 1762.

Sacks, J. (2005). *To heal a fractured world: The ethics of responsibility*. Schocken Books.

Sacks, J. (2016). *Essays on ethics: A weekly reading of the Jewish Bible*. Maggid Books.

Salisbury. (2022). *Sustainability and climate change: Sustainability is our responsibility*. Salisbury. https://www.salisbury.sa.gov.au/services/environment-and-sustainability/sustainability-and-climate-change

Sandelands, L., & Hoffman, A. (2008). Sustainability, faith and the market. *Worldviews, 12*(2–3), 129–145.

Schwartz, E. (1995). Judaism and nature: Theological and moral issues to consider while renegotiating a Jewish relationship to the natural world. *Judaism: A Quarterly Journal of Jewish Life and Thought, 44*(4), 437. +. Gale Resource Literature Center.

Sears, D. (2014). *The vision of Eden: Animal welfare and vegetarianism in Jewish law and mysticism*. Createspace Independent Publisher. https://www.bookdepository.com

Shao, E. (2022, July 8). Biodiversity crisis affects billions who rely on wild species, researchers say. *New York Times*. https://www.nytimes.com/2022/07/08/climate/species-biodiversity-united-nations.html

Smith, A. (1937). *The wealth of nations*. Modern Library.

Smith, H. (1991). *The world's religions*. Harper.

Sonak, S., Sonak, M., & Giriyan, A. (2008). Shipping hazardous waste: Implications for economically developing countries. *International Environmental Agreements: Politics, Law, and Economics, 8*, 143–159. https://doi.org/10.1007/s10784-008-9069-3

Stephen, B., Lee, I., & Kim, J. (2019). *Crashing the climate: How the car industry is driving the climate crisis*. Greenpeace International. https://www.greenpeace.de/publikationen/Crashing%20the%20Climate%20engl%20LF.pdf

Stokstad, E. (2022). Dams are supposed to prevent floods. Some may make them worse. *Science. org.* https://www.science.org/content/article/dams-are-supposed-prevent-floods-some-may-make-them-worse

Tabuchi, H. (2018, June 19). How the Koch brothers are killing public transit projects around the country. *The New York Times*. https://www.nytimes.com/2018/06/19/climate/koch-brothers-public-transit.html

Tessum, C. W., Paolella, D. A., Chambliss, S. E., Apte, J. S., Hill, J. D., & Marshall, J. D. (2021). PM $_{2.5}$ polluters disproportionately and systemically affect people of color in the United States. *Science Advances, 7*(18). https://doi.org/10.1126/sciadv.abf4491

Tieso, I. (2022). Global waste trade-statistics and facts. *Statista*. https://www.statista.com/topics/7943/global-waste-trade/#dossierKeyfigures

Tirosh-Samuelson, H. (2001). Nature in the sources of Judaism. *Daedalus, 130*(4), 99–124.

Tirosh-Samuelson, H. (2019). Judaism responds to the environmental crisis. *The Ecological Citizen, 3*, 41–42.

Tirosh-Samuelson, H. (2022). Judaism and climate change: Environmental ethics and social activism. *Tenenbaum Family Lecture in Judaic Studies*. Emory University (March 8).

Troster, L. (2011). *Ten Jewish teachings on Judaism and the environment*. AJWS On1Foot. https://www.sefaria.org/sheets/114041.1?lang=bi&with=AboutSheet&lang2=en

Troster, L. (2018). Jewish teachings on water. GreenFaith Water Shield.. https://www.faithinwater.org/uploads/4/4/3/0/44307383/jewish_teachings_on_water-greenfaith.pdf

US Global Leadership Coalition. ([USGLC], 2021 March). *Climate change and the developing world: A disproportionate impact*. https://www.usglc.org/media/2021/03/USGLC-fact-sheet-climate-change.pdf

UN Global Compact. (2019). *UN Global Compact – Accenture strategy 2019 CEO study on sustainability: The decade to deliver: A call to business action.* 1–82. https://unglobalcompact.org/library/5715

Union of Concerned Scientists. (2016, February 8). *What's driving deforestation?* ucsusa.org. https://www.ucsusa.org/resources/whats-driving-deforestation

United Nations. (2022). *The sustainable development goals report 2022*. UN Department of Economic and Social Affairs. https://unstats.un.org/sdgs/report/2022/

United Nations Department of Global Communications (2020, May). *Sustainable development goals*. UN. https://sdgs.un.org/goals

United Nations Environmental Programme. (2019, June 7). *How all religious faiths advocate for environmental protection*. https://www.unep.org/news-and-stories/story/how-all-religious-faiths-advocate-environmental-protection

Vogel, D. (2001). How green is Judaism? Exploring Jewish environmental ethics. *Business Ethics Quarterly, 11*(2), 349–363. https://doi.org/10.2307/3857753

Waskow, A. (2000). And the Earth is filled with the breath of life. In A. Waskow (Ed.), *Torah of the Earth: Exploring 4,000 years of ecology in Jewish thought* (Vol. 2, pp. 261–286). Jewish Lights Publishing.

Watson, J. E. M., Shanahan, D. F., Di Marco, M., Allan, J., Laurance, W. F., Sanderson, E. W., Mackey, B., & Venter, O. (2016). Catastrophic declines in wilderness areas undermine global environment targets. *Current Biology, 26*, 2929–2934.

Weintraub, S. H. (2022). *To work and to preserve Judaism and the environment*. Kane Street Synagogue. https://kanestreet.org

Weyler, R. (2018). *A brief history of environmentalism*. Greenpeace.

Wolff, K. A. (2009, December 1). *Bal Tashchit: The Jewish prohibition against needless destruction*. Leiden University. https://hdl.handle.net/1887/14448

Wolff, A., & Neril, Y. (2012). *Torah, trees and caring for the earth*. Jewcology.

Yip, L. (2021). Why marginalized groups are disproportionately affected by climate change. *Earth. org*. https://earth.org/marginalised-grfoups-are-disproportionately-affecgted-by-climate-change/

Yoreh, T. (2019). *Waste not: A Jewish environmental ethic*. State University of New York Press.

Zemer, M. (2003). Ecology as a mitzvah. In W. Jacob & M. Zemer (Eds.), *The environment in Jewish law: Essays and responsa* (pp. 24–33). Berghahn Books.

Chapter 10
A Collective Sustainability Approach Based on the Bahá'í Principles

Legha Momtazian

Abstract Religion and sustainability initiatives have been at the forefront of academic debates with a number of differing perspectives. This chapter contributes to this emerging body of research and proposes a "collective sustainability framework", drawn from the Bahá'í Faith. This framework enables an effective integration of various stakeholders, which benefits both the environment and the society at large, based on the scriptures of the Bahá'í Faith and practices of the Bahá'í community. Such a sustainability framework is built on the principles of justice for all, unity in diversity, unity between science and religion, high station of work, and cessation of war expenditure. The chapter concludes by proposing that such a collective sustainability framework can develop through a "value intersubjectivity" approach, to reflect the context and the environment it applies to. It can be employed in sustainability agendas as well as in organizational policy decisions and sustainability initiatives.

Keywords Bahá'í faith · Sustainability framework · Unity in Diversity · Justice · Peace and Universal Education · Value intersubjectivity

10.1 Introduction

The contribution of religion to development and sustainability initiatives has been the subject matter of academic debates since the last few decades (Taylor et al., 2016; Andregg, 2012; Adriance et al., 2010; Zagonari, 2021; White, 1989). While some scholars have highlighted the positive contributions of religion on existing sustainability discourses (Tomalin et al., 2019), others point to a divisive impact of religions and their myopic focus on the issue. For instance, Andregg (2012, p. 41)

L. Momtazian (✉)
The University of Gloucestershire, Cheltenham, UK
e-mail: lmomtazian@glos.ac.uk

N. Singh et al. (eds.), *Faith Traditions and Sustainability*, Management, Change, Strategy and Positive Leadership, https://doi.org/10.1007/978-3-031-41245-5_10

critiques the conflicts caused by "authoritarian law and militant religion" that creates contestations with science and reason. Rather, he necessitates a "civilizational religion" that could manage the planet and the people on it more humanely, instead of subjugating and exploiting them for short term benefit" (p. 41). On the other hand, Shogar (2011, p. 754) reviews various studies that attribute the recent ecological crises to the contemporary scientific paradigms "dominated by a secularist worldview" and necessitates the importance of spiritual values. Furthermore, Adriance et al. (2010, p. 1) explore the often-overlooked contributions of various religious leaders and their followers and argue that "preserving the earth's biodiversity and fostering human well-being does not have to be a secular undertaking." This debate between the benefits and harms of the contributions of science and religion remains a contentious subject, which has yet to reach a consensus.

In integrating the two schools of thought, Zagonari (2021, p. 5) conducts a multilayered analysis comparing *religious* and *secular* ethics in terms of household determinants of key environmental behaviors. He studied variables such as household waste management, organic food purchases, household energy conservation and ecological footprint. The study concluded that religious and secular ethics complement each other. Religious ethics impact *individual feelings* and contribute to solving local sustainability issues. On the other hand, secular ethics help to solve *collective action problems* and help in achievement of long-run sustainability. Adriance et al. (2010) observe the impact of collaboration between the secular scientific paradigms and religious groups in the case of Appalachian communities. They discuss how pairing various stakeholders including the environmentalists, artists, and religious groups, enhanced the viability, development, and implementation of sustainability programs toward protection of mountain ecosystems.

These multiple sources of evidence and empirical case studies have gradually evolved into local level partnerships between the religious groups and secular activists in collaboration toward sustainability. However, the issue of a "civilizational religion" as phrased by Andregg (2012) still persists, which looks for a unifying set of spiritual principles and institutions that would steer sustainability and development initiatives. This chapter explores the possibility of the contribution of the Bahá'í Faith to this call for a "civilizational religion" as a contemporary world religion, focused on the key tenets of unity in diversity, justice, and betterment of the world as a whole.

This chapter aims to articulate and suggest a generic framework for articulation and agreement on values in sustainable development, in organizations, based on the Bahá'í principles. It aims to enable the enactment and measurement of these values, as a potential inclusive, peace-oriented value system, established on the principles of oneness of humanity and focused on betterment of the world (with a collective approach). These principles are explored in detail, in the next section.

The chapter begins by analyzing the relationship between religious values and sustainability agendas in literature and then goes on to highlight the principles from the Bahá'í scripture and the novel perspective on sustainability they bring to existing literature. The chapter concludes with an analysis and discussion of ground-level initiatives informed by Bahá'í principles.

10.2 Religious Values and Sustainability Agendas

To enable effective partnership between religions and sustainability, some studies (Ives & Kidwell, 2019; Narayanan, 2013; Burford et al., 2013) have focused on existing "sustainability agendas," which often lack any trace of religions and their principles or keep them to a minimum. Burford et al. (2013), however, argue that inclusion of religious principles and doctrines can enable the inclusion of "humane values" as the fourth pillar of sustainable development, along with existing dimensions of environmental, economic, and social sustainability. Values, here, are intended as "principles or standards of behavior" and "people's judgments about what is important in life" (Burford et al., 2013, p. 3038). They evaluated the presence of "values" in existing sustainability agendas and found that the Commission for Sustainable Development's (CSD) millennium declaration incorporated the importance of values as "rooted principles," which steer integrated development (Burford et al., 2013, p. 3039).

Despite this recognition, the final report of the Rio +20, the United Nations' Conference on Sustainable Development, missed out on dealing with ethics and values altogether. It is notable that later, in 2015, the United Nations' summit in New York, which resulted in 2030 Agenda for Sustainable developments, includes a variety of values including "values of peace, dialogue and international cooperation" (General Assembly, 2015, p. 12), diversity, physical and mental health, and economic and environmental advancements as some of its main goals. This indicates a significant improvement in comparison to previous sustainability agendas and sets the motion for a diverse world with collaborative citizenship. In relation to the goals focused on diversity, in Declaration No. 36 it says:

> We pledge to foster intercultural **understanding, tolerance, mutual respect and an ethic of global citizenship and shared responsibility**. We acknowledge the natural and cultural diversity of the world and recognize that all cultures and civilizations can contribute to, and are crucial enablers of, sustainable development. (General Assembly, 2015, p. 10)

The stress on diversity and its inclusion in the agenda is significant in providing a conducive platform for integration of secular and religious (various religions') involvements, as well as broader variety of perspectives and experiences of multiple stakeholders. However, this constitutes only an initial step.

Singh and Clark (2016, p. 271) translate the focus of the 2030 agenda on the five Ps, namely, "planet, people, prosperity, peace, and partnership," to the views of ten different religions and their potential contributions. They include Christianity, Judaism, Islam, Bahá'í Faith, Hinduism, Sikhism, Buddhism, Confucianism, Daoism, and Lenape. These approaches, while having many similarities, offer differing perspectives and strategies, which is not surprising, since values are not universally defined and characterized in research and practice.

The inability to produce objective measures to evaluate values and their fulfilment has been considered as a barrier to the inclusion of values in previous sustainability agendas. Hence, to enable development of a framework for inclusion of values and collaboration between religious and secular efforts, a universal understanding of values and their implications is deemed necessary. To overcome this barrier, Burford et al. (2013) discuss "value intersubjectivity" to collectively conceptualize values through mutual agreement and define and measure them within specific contexts, which can, consequently, enable adequate definition and evaluation of values in sustainable development agenda. To enable value intersubjectivity, they suggest "a polycentric approach, based on the creation of multiple, diverse, peer-elicited, indicator sets and assessment tools" and "reference sets of indicators and survey questionnaires which can be used in their standard forms to generate national-level statistics but can also be tailored to local contexts" (p. 3050). They highlight that it requires a "transdisciplinary approach" in terms of "facilitating the deep participation of non-scientific stakeholders in research and the —direct application of scientific knowledge in both political decision-making and societal problem-solving" (Burford et al., 2013, p. 3053). This approach can enable adequate definition, inclusion, and evaluation of values in sustainable development agenda as well as organizational sustainability initiatives, through creation of indicator sets and assessment tools in each conceptual, geographical, and organizational context.

Taking this approach to action, in relation to including religious followers and institutions as part of the "nonscientific stakeholders," Taylor et al. (2016, p. 354) pose a question:

> What are the signs and possibilities of an emerging civil earth religion or planetary civilization in which people with very different metaphysical understandings could find common ground, recast themselves as citizens of the Earth (whatever other identities they may hold), and sincerely work together to create sustainable and equitable societies?

To answer this question in the context of a "civilizational religion," this chapter explores the principles of the Bahá'í Faith as a contemporary world religion, established in 1844 in Iran. Baha'i Faith aims to establish unity in diversity and prosperity of the world population ('Abdu'l-Bahá, 1998). These principles include universal peace (with inclusion of members from all countries), availability of and access to education for all, unity of religions and recognizing their common purpose toward unity of humankind, equality of men and women, and a stop to war expenditures (Shoghi Effendi, 1991b, p. 203). This set of principles exhibit an attempt toward world citizenship with practical guidance on their achievements, including a path to this universal peace. For instance, elaborating on one of such steps on this path, these principles encourage establishment of a universal language. Shoghi Effendi (1991b, p. 203) explains that in a meeting with representatives from all nations and cultures, either a language be selected among the existing languages or one to be developed, in order that all people of the world learn the common language along with their mother tongues, so that the concept of stranger is gradually removed, in a global culture. This is one of such principles that can inform UN sustainability agenda toward a collective peace-oriented approach to sustainability.

10.3 The Thesis of the Bahá'í Faith and Recommendation for Inclusion of Values in the Agenda

The Bahá'í Faith is a contemporary world religion, with a monistic approach to existence of one God and progressive manifestations throughout time, as different stages in the history of one religion, while believing that the revelation of Bahá'u'lláh (1817–1892) is the most recent (not the last) one (Smith, 2000). The principles of the Bahá'í Faith are centered around unity and oneness of humankind; and the collection of values/virtues, the guiding principles and instructions, and its administrative system operate according to the same approach, while valuing human dignity and choice. For instance, 'Abdu'l-Bahá, 1982b, pp. 299–300) introduces the unity of religions and states that religion "should be the cause of fellowship," but if it causes "prejudice and enmity," then "it is fruitless." Similarly, followers are guided by Bahá'u'lláh (1992, p. 20), the founder of the Bahá'í Faith, as follows: "Observe My commandments, for the love of My beauty." Similarly, the administrative structure of the Bahá'í Faith does not have clergy and is elective, and the elective body is advised to take consult from the members of the community and serve with purity of motive, rather than taking an authoritative approach to protection and expansion of the religion (Shoghi Effendi, 1991a, 1970).

In this paragraph, I elucidate upon two main aspects of the Bahá'í Faith that contribute to the debate on sustainability, through the lens of the "civilizational religion." One of these is the unity of religions, which takes a unifying approach to sharing values and appreciation of diversity, hence overcoming the divisive impact previously attributed to the contributions of religions. The second is the lack of clergy and the elective nature of the administrative body of the Bahá'í Faith, which establishes and encourages the spirit of consultation and collective articulation of values, their applications, and personal assessment as opposed to clerical approval of religiosity. The nature and experience of religion in this manner can create a platform for establishment of "value intersubjectivity" as addressed by Burford et al. (2013) and co-definition of the organizational value systems and their evaluation, collectively.

The following sections discuss the Bahá'í principals that have been found to have a direct impact on a sustainability approach. They include justice, unity in diversity, the value of work, and cease to war expenditure.

10.3.1 Justice Through Moderation and Share of Wealth

In Bahá'í Faith, individual responsibility is delegated to act with care and moderation. 'Abdu'l-Bahá (1972, p. 153) states: "If it be right for a capitalist to possess a large fortune, it is equally just that his workman should have a sufficient means of existence." So, the owners of capital are advised to be "moderate" in "withdrawal of profit" and concentrate on "the welfare of the poor." Hence, they are encouraged to share their profit with the workmen and pay them enough salaries that they can save for their time of need ('Abdu'l-Bahá, 1990a, pp. 273–277). This value is unique in its focus and approach in relation to both religious and secular value systems. While looking after the poor has commonly been considered as a moral responsibility of the rich, in both religious and secular ethics, it is often rooted in charitable contributions and development activities. The above principle, however, encourages the owners of capital to undertake moderate and responsible withdrawal of profit as a part of their personal and social responsibility.

Accordingly, in relation to sustainability, this principle discourages excess in accumulation of wealth as well as its consumption (White, 1989). The latter is also advocated by many sustainability researchers (Koh & Lee, 2012; Assadourian, 2010; Mroz, 2010) on the grounds that it promotes sustainable consumption (Kurenlahti & Salonen, 2018). In recent years, there have been numerous references to ethical consumerism (Gillani & Kutaula, 2018) or sustainable consumption in the British press in their coverage of sustainability issues as well (Diprose et al., 2018). The emphasis on ethical and sustainable consumerism, similarly, discourages excess consumerism and indirectly points to moderation, although with a different justification. Notably, while excess accumulation and consumption of wealth is discouraged in existing sustainability literature, the responsibility of the owners of capital in their withdrawal of profit is unique to the Bahá'í principles and points to a level of responsibility that is deeply rooted in individuals' consciousness and sacrifice and care for the welfare of the society, which is different from contributing excess profit to humanitarian causes.

10.3.2 Justice Through Equitable Distribution of Natural Resources

Shoghi Effendi (1991b, p. 203) discusses establishment of a "world commonwealth" that unites "all nations, races, creeds and classes," while autonomy and freedom of the state members are maintained. Within the framework of a world commonwealth, he introduces a world legislature, a world executive, a world tribunal, and a world federal system, members of which are representatives of all nations and aim to safeguard and satisfy the needs of all the nations. In such a commonwealth, Shoghi Effendi (1991b, p. 203) states that the world legislature would "ultimately control the entire resources of all the component nations, and will enact such laws as shall be required to regulate the life, satisfy the needs and adjust the relationships of all races and peoples"; and the world federal system would have control and "authority over its unimaginably vast resources," would blend and embody "the ideals of both the East and the West, liberated from the curse of war and its miseries," and would be "bent on the exploitation of all the available sources of energy on the surface of the planet." This commonwealth, being made up of all nations/countries and serving all nations/countries equitably, is in contrast to the operation of the

United Nations (UN), the Conference of the parties (COP), the European Union (EU) in their very (membership) nature, being made up of member-countries and serving the interests of the member-countries, consequently, negating the principles of a universal social justice and discriminating the non-member nations.

In terms of equitable distribution of natural resources in a world commonwealth mentioned above, sustainability literature shows a great concern for their exploitation and distribution (Kuhlman & Farrington, 2010; Agarwal, 1989). For instance, the impact of globalization on providing international access to natural resources has been challenged. Tu et al. (2019) explore the resilience of the system when multiple users harvest the same resources. They explain that a positive or negative impact of globalization depends on the interconnectedness of the network and its modularity. Hence, they recommend reciprocated trade (Tu et al., 2019). They state that major importers of food often only export small amounts of food, to a small number of countries, which indicates the "inequality in the global distribution and development of food resources relative to the population of each country" and decreases the resilience and sustainability of the system (Tu et al., 2019, p. 288). This discussion highlights the significance of a resilient and reciprocal system in access to national resources. However, it is different from the thesis of a "world commonwealth" as above, which proposes equitable access to and distribution of natural resources, in the interest of all countries, as the property of the commonwealth. This can contribute to addressing the issues of neglect/destruction of natural resources as well as equitable access, as ongoing causes of poverty (Addae-Korankye, 2014; Anderson et al., 2006).

From a different perspective, Ragnarsdóttir et al. (2012) employ systems analysis to assess the long-term sustainability of global supply and consumption of natural resources. They find that despite the recent technological advances in supply of resources, the inefficient consumption of energies and natural resources such as gas and oil are leading to unsustainability.

Accordingly, establishment of a reciprocal, responsible, and resilient system of access, distribution, protection, exploitation, and consumption of national resources in a world commonwealth is proposed in Bahá'í scripture, where individual responsibility and moderate consumption (Bahá'u'lláh, 1988) are coupled with equitable sharing of the resources ('Abdu'l-Bahá, 1990a, pp. 273–277), so that a collective justice can enable sustainability.

10.3.3 Unity Between Science and Religion

Another principle of the Bahá'í Faith in relation to sustainability is the consideration of a unity between science and religion, through which the two forces join in the service of the entire world (Shoghi Effendi, 1991b, p. 205). Technological and scientific advances have enabled mass production, development of the quality of crops and raw material (Singh et al., 2020; Pirzada et al., 2020), and reducing carbon emissions (Taheri-Najafabadi, 2013). However, responsible employment of technology is a challenge to sustainability movements, as production continues to be dictated by business interests. Fuchs (2008, p. 291) argues that achieving a sustainable society "demands a conscious reduction of profits by not investing in the future of capital, but the future of humans, society, and nature."

This unity between science and religion adds another dimension to the "purpose orientation" and objectivity of philosophy of science (John, 2021), subjecting it to human dignity and well-being, morality, peace, and protection of "all" humankind and all planet citizens. This unity can, thus, channel the development and employment of purely business and commercial prerogatives of environmental initiatives, away from "value-free" science (John, 2021, p. 20), and embed them in a pro-environment framework.

10.3.4 Diversity

The principle of unity in diversity considers the universe as a greater whole, which requires appreciation and flourishing of all aspects of its diversity, to enable and protect the sustainability of its peace and growth. It focuses on access to opportunities for all ('Abdu'l-Bahá, 1982a, p. 29) resulting in an enhanced positive outcome, the same way that the "vari-colored flowers of one garden" enhance the beauty of the garden ('Abdu'l-Bahá, 1998, p. 25). Thus, unity and peace according to Bahá'í principles are interpreted as "unity in diversity."

It is worth nothing that diversity as addressed by scholars of sustainability or ecology is often focused on biodiversity (Díaz et al., 2020; Bennett et al., 2015; Adriance et al., 2010; Bishop, 1993) referring to "the variety of life" or "collectively to variation at all levels of biological organisation" (Gaston & Spicer, 2013, p. 16). In Bahá'í scripture, diversity refers to the variety of "races, tribes, and communities of the world," but it further includes the "different and varied customs, habits, tastes, character, inclinations and ideas" ('Abdu'l-Bahá, 1982b, p. 290). This articulation extends the definition of diversity to inclusion of habits, characters, and opinions as well, which goes beyond the dictates of nature, toward the evolving nature of diversity, including mental capabilities and psychological conscience (including ideas, thoughts, interpretations, meanings, fantasies, and values) often interpreted as "soul" or "spirit" (Giegerich, 2020). Such diversity, as stressed in Bahá'í scripture, requires systems and mechanisms to flourish and bear fruits ('Abdu'l-Bahá, 1998, p. 25). Hence, in discussing diversity based on Bahá'í scripture, this chapter takes the broader approach to diversity, beyond biodiversity, and conceptualizes it as an essential precondition in sustainability planning and implementation, enabling all components of nature and cultures to foster and evolve.

10.3.5 Cessation of War and War Expenditure

Bahá'í scripture strongly discourage war and spending on warfare expenditure due to the loss of life, grief, destructions, and extreme financial expenses it causes ('Abdu'l-Bahá, 1979, p. 10), which is supported by several development and sustainability studies as well (Smith, 1980; Deger & Smith, 1983). For instance, the US military expenditure in 2020 amounted to 766.58 billion US dollars (Duffin, 2022), and the UK defense spending in 2020/2021 amounted to 42.4 billion British pounds (Kirk-Wade, 2022). However, despite the amount of financial costs and loss of environmental, structural, and infrastructural resources, this factor has not gained a critical significance in sustainability debate.

Moreover, Bahá'í scripture discourages nationalism and promote a global outlook. Bahá'u'lláh (1988, p. 167) says: "It is not for him to pride himself who loveth his own country, but rather for him who loveth the whole world. The earth is but one country, and mankind its citizens." This principle further supports the approach of the Bahá'í Faith to establishment of peace and unity as a means of sustainable citizenship. The debate on global citizenship versus national citizenship is still ongoing in media, political speeches, and in research, with opposing opinions (Buckner & Russell, 2013; Harshman, 2018). Zagonari (2021, pp. 11–12) aims to close this gap and states that global citizenship "does not require a single global ideology or a homogeneous world culture" and recognizes "religious or secular ethics, which differ among cultures, societies, and economies," an approach that further applies to organizational sustainability initiatives as well.

However, there are also contrary views on the idea of "global citizenship." For instance, Bowden (2003) argues against global citizenship on the basis that it causes "statelessness." He compares it to the condition of refugees, with no access to institutions one would belong to, as well as the lack of cultural pluralism in cosmopolitan perspective of global citizenship. Therefore, Bowden (2003, p. 360) argues that being a global citizen is a claim that only "modern, affluent global bourgeoisie" can make. However, the above articulation of being the citizen of the world is put forward based on a system of values and guidelines that aim for establishment of universal peace along with its supporting means and mechanisms (Shoghi Effendi, 1991b). These would be inclusive of all nations and tribes and aim to create equitable access for "all," especially the disadvantaged sections of society. In such a system, the army is not a means of conflict resolution, and a world tribunal would undertake a different role compared to that of a traditional army. This would take place in the broader framework of a "world commonwealth" (Shoghi Effendi, 1991b, p. 203).

10.3.6 High Station of the Work to Benefit Self, Others, and the World

Work has been regarded as worship in Bahá'í scripture, if done in the "spirit of service" ('Abdu'l-Bahá, 1998, p. 83). Hence, believers in the Bahá'í Faith are directed to avoid "idleness and sloth" and "engage in some form of occupation, such as crafts, trades and the like," which would benefit them and others (Bahá'u'lláh, 1988, p. 26) and "to serve the general good" ('Abdu'l-Bahá, 1990b, p. 103). Hence, Bahá'ís consider engagement in activities that cast harm on the world as prohibited, such as serving alcohol in their premises (Bahá'í Forums, 2013), engagement in dealing drugs and in the armed forces, or similar activities that threaten human dignity or their physical, mental, or spiritual well-being. In sustainability debates, existence of and access to jobs is considered as an issue that has led to poverty and challenged sustainability for a long time, and governments constantly aim to include them in their development objectives. For instance, this concern is evidenced in UK sustainability appraisal, which includes "jobs, economic growth, housing, transport, services" and various aspects of the rural and urban environment (George, 2001, p. 101). However, they are still on their long way to solving the unemployment problem. For instance, in the UK, the unemployment rate in 2022 is 3.8% of the population over 64, while employment rate of 16–64 is 75.5% (Office for National Statistics, 2022). This encouragement to engage in some form of occupation in Bahá'í scripture looks at the micro-individual aspect of access to jobs, which is further coupled with availability of job from the macro-perspective as the responsibility of the governments. Here, individuals are advised to maintain their religious service and devotion "along" with their occupation, as opposed to devoting all their time to the service of the Faith, which highlights the value of engaging in a profession in this approach, along with discouragement of "priesthood" and leading a single-sided life (Shoghi Effendi, 1988, p. 520). This approach to discouraging complete devotion of time to religious work is unique to this scripture, and encouragement of all individuals to engage in a profession can directly impact development and sustainability.

10.4 Value Intersubjectivity and Sustainability Agenda

The above discussions provide an interpretation of the Bahá'í principles in relation to sustainability and development. Considering these principles and the argument of "value intersubjectivity" by Burford et al. (2013) as discussed in previous sections, this section aims to suggest a framework to enable inclusion and development of values in development and sustainability agenda. In order for the agenda to be "based on" the framework of values, this chapter suggests that the value framework is defined, as the initial step, and then the agenda and goals are decided and planned accordingly, "established" based on the value framework.

The study by Burford et al. (2013, p. 3040) summarized "16 most frequent global values included in 22 international documents" as equality, responsibility, participation, cooperation, dignity, freedom, security, peace, protect, respect, dialogue, integrity, diversity, tolerance, justice, and solidarity. While the values identified based on the analysis of the Bahá'í writings support the above values, they expand the concepts and applications of the values that have not been considered before and further offer a focal point for employment of the values, objectives, plans, and efforts in a cohesive system. These values include a) "justice for all" through moderation in their withdrawals by the owners of capital; b) sharing the benefits of the company with the workers; c) equitable access to, and distribution and protection of natural resources for all through operation of a world commonwealth; d) "unity in diversity" through appreciation (rather than management) of diversity; e) respecting the high station of human and avoidance of consumption of drugs and alcohol, and instead, engagement in a profession and self-development; f) unity between science and religion; and g) cease to war expenditure.

The two core values of "unity in diversity" and "justice for all" operate as the focal points of this framework. Accordingly, this chapter suggests the following framework as a forum to be agreed upon, or changed as needed, at the initial step before developing any sustainability agenda.

Value framework, fostering unity in diversity and justice for all, includes:

- Values concerned with human dignity and well-being, supported by the establishment of a world commonwealth (equality and appreciation of diversity, universal education, work, physical and mental health and spiritual well-being, justice, and fairness for all (including the vulnerable population), prevention of violence, promotion of human rights, and equitable access to natural resources)
- Values concerned with global prosperity, supported by the establishment of a world commonwealth (peace, poverty alleviation, standard of living, shared responsibility and benefits, mutual care, service of humanity, moving beyond self and personal fulfilment, appreciation of diversity beyond biodiversity), a world tribunal and a world federal system that safeguards and satisfies the needs of all the nations, global monetary system, and global language
- Values concerned with the earth protection, supported by the establishment of a world commonwealth (which protects the nature and natural resources mindfully):
  - With animals' protection
  - With plants protection
  - With sea protection

This framework has been summarized in Fig. 10.1, which proposes establishment of a world commonwealth, or an organizational commonwealth, to enable intersubjectivity in development and agreement of a value system at the levels of earth protection, global prosperity, and human dignity and well-being.

Some of the virtues that underlie care and concern for the environment and the entire world include "tolerance, compassion, trustworthiness, kindness, willingness

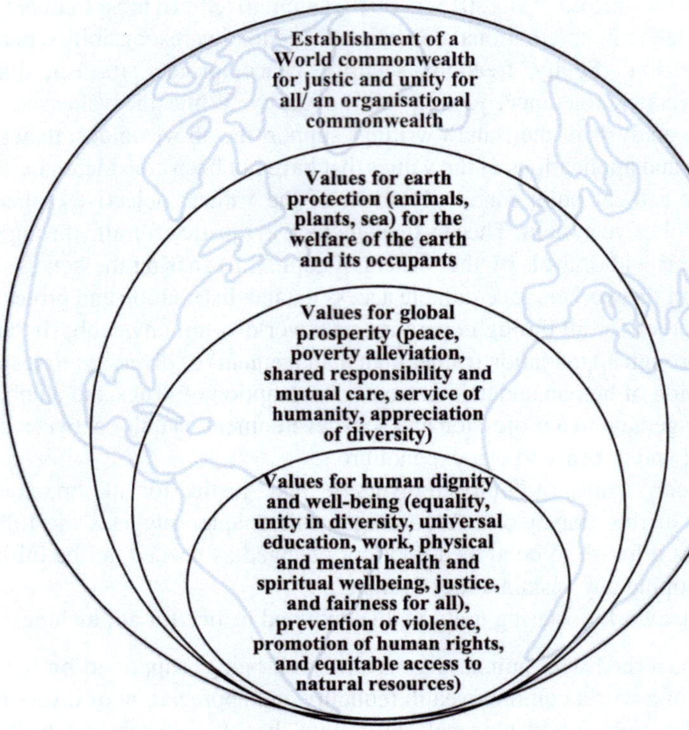

Fig. 10.1 The value framework for development of sustainability agenda and organizational value systems

to sacrifice for the common good, humility" (Kavelin, 2008), which would require individual effort as well as formal and informal education to develop and enable a world citizenship approach. In terms of education, 'Abdu'l-Bahá (1990a, pp. 8–9) articulates three kinds, namely, "material, human, and spiritual," where "material education is concerned with the progress and development of the body," "human education signifies civilization and progress," and "divine education is that of the Kingdom of God." He signifies all three aspects of education and encourages individuals to develop all aspects of their lives accordingly.

Education of values play a significant role in this framework, in terms of organizational initiatives toward sustainability as well as national and international programs. Burford et al. (2013, p. 3038) distinguish between the concept of values as standards and principles that guide our behaviors and "values as one's judgment of what is important in life." To enable instilment of each of these interpretations of values in education, it requires their definition and measurement mechanism, which could be facilitated by "value intersubjectivity," through clarifying and specifying values in the contexts and with participation of involved members. This methodology has not been employed in formal education of values before and is recommended for further research. However, there are several case studies that have explored the involvement of the Bahá'í community in relation to spiritual education and sustainability, in various contexts and with various approaches, in their attempt to fulfil the principle of "universal education" as one of the core precepts of the Bahá'í Faith (Bahá'u'lláh, 1992, p. 162). These case studies are discussed below to enable refinement and conclusion of the discussions of this chapter.

10.5 Grassroot Level Initiatives Influenced by Bahá'í Principles

One example is Podger's (2009) critical analysis approach in the study of the contribution of the American Bahá'í Community to education for sustainability (EfS), based on their focus on the oneness of humankind and development of spiritual qualities to enable individuals to offer service to society (as an outcome). These contributions that had been studied included partnership of the National Spiritual Assembly (NSA) of the Bahá'ís of the United States, the national elected administrative body of the Bahá'í Faith, in the US Partnership for Education for Sustainable Development (USPESD) and its impact on the USPESD's efforts and plans. Podger analyzes the contributions to EfS in three themes of expansion of consciousness and identity to activate the potentiality for action, recognition of humanity's spiritual nature and development of spiritual capabilities, and development of a service orientation to life. Podger (2009, pp. 67, 72) concludes that spirituality (as competencies of heads, hearts, hands, spirits) is found essential in the EfS, particularly in education of the children "toward becoming citizens capable of designing and maintaining sustainable societies" and in "motivating individuals and communities to engage in transformation towards sustainability."

This case study indicates that sustainability requires long-term and short-term education of values among children and all members of society respectively, to enable employment of "unity in diversity" as a guiding value to motivate individual and organizational contributions to sustainability.

Another case that comprises the impact of faith-based communities and organizations on the ecosystem services in Oceania (a geographical region including Australasia, Melanesia, Micronesia, and Polynesia) has been explored. Here, some of the activities conducted by the Bahá'í community include planting indigenous trees on all the islands as part of the "Breath of Life" tree planting campaign; ongoing initiatives that focus on conservation education and training; projects to protect and restore the environment; use of the arts to inspire active commitment to environmental protection and development; and "advocacy for sustainable development at local, national, and international levels" (Rowland, 2019, p. 4).

The paper concludes that the Bahá'í community recognizes "humans as the stewards of the natural world" and views biodiversity as a blessing, "with science and religion as complementary aspects of human progress" (Rowland, 2019, p. 4).

Bahá'í perspective is a "combination of this organic understanding of the co-evolution of the spiritual and the material combined with a consciousness of the oneness of humanity" (Kavelin, 2008, p. 74). In the Encyclopaedia of Peace Education by Teachers College, Columbia University, Gervais (2008, pp. 1–2) lists a variety of Bahá'í initiatives on peace education, such as City Montessori School in Lucknow, Northern India, that was awarded the UNESCO 2002 Prize for Peace Education, and Bahá'í International Community (BIC)'s contributions to the UN, which was recognized by the United Nations as an international nongovernmental organization in 1948 and was granted consultative status in 1970 with ECOSOC and in 1976 with UNICEF.

These examples indicate that sustainability initiatives of the Bahá'í community go beyond environmental challenges to an overall objective of betterment of the world. These cases highlight the value of peace and unity and justice in sustainability initiatives of the Bahá'í community and further contributes to the value framework discussed in the previous section, with an emphasis on moral/spiritual education.

10.6 Conclusions

This chapter explored the approach of the Bahá'í Faith to sustainability and aimed to contribute to development of a value framework to operate as a starting point for intersubjective agreement on the initial values in an organizational context as well as providing a basis for sustainability agendas. Accordingly, it was discussed that the core principles concerning environmental and sustainability initiatives of the Bahá'í Faith are the principles of "unity in diversity" and "justice for all," which enable care, compassion, and moderation, in desires, actions, and concern for others. In other words, it was found that sustainability, unity, and justice were interwoven, and its establishment and progress were dependent on both values.

The Bahá'í Faith further considers a unity between science and religion, through which one joins forces with the other in the service of the entire world. These principles consider the universe as a greater whole, where diversity is an integral part of, and its appreciation and flourishing enable the fulfilment of its full potential. At the same time, they argue that the natural resources in each country do not belong to a nation but to the whole world, and the benefits therefrom need to be shared among all citizens of the world equitably. This was discussed as a pillar in establishment of a world commonwealth, with members from all the nations, and its legislation, executive, and tribunal boards would be representatives of all nations and protective of their rights.

Accordingly, in this approach, sustainability arises as a case of cooperation, understanding, and appreciation and evolvement of unity in diversity, in all social and economic systems. It also considers the processes and structures that may cast harm to the body or the soul of the environment, including physical harm, or social and emotional harm, to human beings or other habitants of the globe. It, thus, calls for every individual's conscience and efforts, to see themselves as a member of the body of the world, where if one part is hurt the whole body would be at unease, and all the mechanisms would endeavor to resolve the issue, in a collective effort. Hence, a value framework based on the Bahá'í principles was conceptualized in this chapter and was focused on promotion of well-being for all, with responsible and reciprocal participation of all individuals and societies. Such a framework would require representatives of all member nations to discuss, define, and specify the practical aspects of the values, agree on the developed framework, and develop the sustainability agenda based upon.

In an organizational context, this would interpret, in a smaller scale, to a platform in every organization to enable conversations about the organizational values established on unity, justice, and sustainability and to develop a collaborative organizational framework that enables well-being and growth of the employees, the environment, and the wider society, where the organization is involved with and is taking responsibility for.

References

'Abdu'l-Bahá. (1972). *Paris talks* (11th Ed.). UK Bahá'í Publishing Trust, reprint.

'Abdu'l-Bahá. (1979). *Foundations of world unity.* US Baha'i Publishing Trust.

'Abdu'l-Bahá. (1982a). *'Abdu'l-Bahá in London.* UK Baha'i Publishing Trust.

'Abdu'l-Bahá. (1982b). *Selections from the writings of 'Abdu'l-Bahá.* Baha'i World Centre.

'Abdu'l-Bahá. (1990a). *Some answered questions.* US Baha'i Publishing Trust.

'Abdu'l-Bahá. (1990b). *The secret of divine civilization.* US Baha'i Publishing Trust.

'Abdu'l-Bahá. (1998). *'Abdu'l-Bahá on divine philosophy.* (E. Frazer Chamberlain, Ed.) Tudor Press.

Addae-Korankye, A. (2014). Causes of poverty in Africa: A review of literature. *American International Journal of Social Science, 3*(7), 147–153.

Adriance, P., Aiken, W., Brown, P., Carmichael, C., Coddington, N. D., et al. (2010). Religion in a sustainable world: What is the role of religious institutions in fostering ecological sustainability? *Ecopsychology, 2*(1), 5–11.

Agarwal, B. (1989). Rural women, poverty and natural resources: Sustenance, sustainability and struggle for change. *Economic and Political Weekly*, pp. WS46–WS65.

Anderson, J., Benjamin, C., Campbell, B., & Tiveau, D. (2006). Forests, poverty and equity in Africa: New perspectives on policy and practice. *International Forestry Review, 8*(1), 44–53.

Andregg, M. M. (2012). Religion for a sustainable civilization. *Dialogue and Universalism, 22*(2), 41–54.

Assadourian, E. (2010). Transforming cultures: From consumerism to sustainability. *Journal of Macromarketing, 30*(2), 186–191.

Bahá'í Forums. (2013, November 12). *What professions can a Baha'i work as?* Retrieved 08 2022, from Bahá'í Forums: https://bahaiforums.com/t/what-professions-can-a-bahai-work-as.9916/

Bahá'u'lláh. (1988). *Tablets of Bahá'u'lláh Revealed After the Kitáb-i-Aqdas* (Vols. pocket-size edition). US Bahá'í Publishing Trust.

Bahá'u'lláh. (1992). *The Kitab-i-Aqdas.* Baha'i World Centre.

Bennett, E., Cramer, W., Begossi, A., Cundill, G., Díaz, S., Egoh, B., et al. (2015). Linking biodiversity, ecosystem services, and human well-being: Three challenges for designing research for sustainability. *Current Opinion in Environmental Sustainability, 14*(1), 76–85.

Bishop, R. (1993). Economic efficiency, sustainability, and biodiversity. *Ambio, 22(2/3)*, 69–73.

Bowden, B. (2003). The perils of global citizenship. *Citizenship Studies, 7*(3), 349–362.
Buckner, E., & Russell, S. (2013). Portraying the global: Cross-national trends in textbooks' portrayal of globalization and global citizenship. *International Studies Quarterly, 57*(4), 738–750.
Burford, G., Hoover, E., Velasco, I., Janoušková, S., Jimenez, A., Piggot, G., et al. (2013). Bringing the "missing pillar" into sustainable development goals: Towards intersubjective values-based indicators. *Sustainability, 5*(7), 3035–3059.
Deger, S., & Smith, R. (1983). Military expenditure and growth in less developed countries. *Journal of Conflict Resolution, 27*(2), 335–353.
Díaz, S., Zafra-Calvo, N., Purvis, A., Verburg, P., Obura, D., Leadley, P., et al. (2020). Set ambitious goals for biodiversity and sustainability. *Science, 370*(6515), 411–413.
Diprose, K., Fern, R., Vanderbeck, R., Chen, L., Valentine, G., Liu, C., & McQuaid, K. (2018). Corporations, consumerism and culpability: Sustainability in the British press. *Environmental Communication, 12*(5), 672–685.
Duffin, E. (2022, June 21). *U.S. military spending from 2000 to 2020.* Retrieved 08 2022, from Statistica.com: https://www.statista.com/statistics/272473/us-military-spending-from-2000-to-2012/#:~:text=In%202020%2C%20the%20United%20States,when%20adjusted%20to%20 2019%20dollars
Fuchs, C. (2008). The implications of new information and communication technologies for sustainability. *Environment, Development and Sustainability, 10*(3), 291–309.
Gaston, K., & Spicer, J. (2013). *Biodiversity: An introduction*. Wiley.
General Assembly. (2015, September 15). *Resolution adopted by the general Assembly on 11 September 2015. A/RES/69/315 15 September 2015. New York: United Nations.* Retrieved 07 2022, from un.org: https://documents-dds-ny.un.org/doc/UNDOC/GEN/N15/291/89/PDF/ N1529189.pdf?OpenElement
George, C. (2001). Sustainability appraisal for sustainable development: Integrating everything from jobs to climate change. *Impact Assessment and Project Appraisal, 19*(2), 95–106.
Gervais, M. (2008). Baha'i faith and peace education. In *Encyclopedia of peace education* (pp. 1–6). Teachers College, Columbia University.
Giegerich, W. (2020). *What is soul?* Routledge.
Gillani, A., & Kutaula, S. (2018). An introduction to special issue: Sustainability and ethical consumerism. *Management Decision, 56*(3), 511–514.
Harshman, J. (2018). Developing global citizenship through critical media literacy in the social studies. *The Journal of Social Studies Research, 42*(2), 107–117.
Ives, C., & Kidwell, J. (2019). Religion and social values for sustainability. *Sustainability Science, 14*(5), 1355–1362.
John, S. (2021). *Objectivity in science*. Cambridge University Press.
Kavelin, C. (2008). An individual Bahá'í perspective on spiritual aspects of cultural diversity and sustainable development: Towards a second enlightenment. *International Journal of Diversity in Organisations, 8*(1), 71–79.
Kirk-Wade, E. (2022, April 06). *UK defence expenditure.* Retrieved 08 2022, from House of Commons Library: https://commonslibrary.parliament.uk/research-briefings/ cbp-8175/#:~:text=In%202020%2F21%2C%20defence%20spending,increase%20of%20 %C2%A31.7%20billion
Koh, L., & Lee, T. (2012). Sensible consumerism for environmental sustainability. *Biological Conservation, 151*(1), 3–6.
Kuhlman, T., & Farrington, J. (2010). What is sustainability? *Sustainability, 2*(11), 3436–3448.
Kurenlahti, M., & Salonen, A. (2018). Rethinking consumerism from the perspective of religion. *Sustainability, 10*(7), 1–18.
Mroz, B. (2010). Consumerism vs. sustainability: The emergence of new consumer trends in Poland. *International Journal of Economic Policy in Emerging Economies, 3*(1), 1–15.
Narayanan, Y. (2013). Religion and sustainable development: Analysing the connections. *Sustainable Development, 21*(2), 131–139.

Office for National Statistics. (2022, August 16). *Unemployment*. Retrieved 08 2022, from ONS.gov.uk: https://www.ons.gov.uk/employmentandlabourmarket/peoplenotinwork/unemployment#timeseries
Pirzada, T., de Farias, B., Mathew, R., Guenther, R., Byrd, M., Sit, T., et al. (2020). Recent advances in biodegradable matrices for active ingredient release in crop protection: Towards attaining sustainability in agriculture. *Current Opinion in Colloid & Interface Science, 48*(1), 121–136.
Podger, D. (2009). Contributions of the American Bahá'í community to education for sustainability. *Journal of Education for Sustainable Development, 3*(1), 65–74.
Ragnarsdóttir, K., Sverdrup, H., & Koca, D. (2012). Assessing long term sustainability of global supply of natural resources and materials. In C. Ghenai (Ed.), *Sustainable development-energy, engineering and technologies-manufacturing and environment*. InTech.
Rowland, P. (2019). Ecosystem services and faith communities in Oceania. *Ecosystem Services, 39*, 1–6.
Shogar, I. (2011). Reconstruction of the worldview as strategy for environmental survival and sustainability. In H. Gökçekus, U. Türker, & J. W. LaMoreaux (Eds.), *Survival and sustainability* (pp. 751–758). Springer.
Shoghi Effendi. (1970). *Dawn of a new day*. Bahá'í Publishing Trust of India.
Shoghi Effendi. (1988).*Lights of Guidance*: A Bahá'í Reference File. (H. B. Hornby, Ed.) Bahá'í publishing trust.
Shoghi Effendi. (1991a). *Compilation of compilations* (Vol. II). Baha'i Publications Australia.
Shoghi Effendi. (1991b). *The world order of Baha'u'llah*. US Bahá'í Publishing Trust.
Singh, K., & Clark, J. (Eds.). (2016). Voices from religions on sustainable development.
Singh, B., Trivedi, P., Egidi, E., Macdonald, C., & Delgado-Baquerizo, M. (2020). Crop microbiome and sustainable agriculture. *Nature Reviews Microbiology, 18*(11), 601–602.
Smith, R. (1980). Military expenditure and investment in OECD countries, 1954–1973. *Journal of Comparative Economics, 4*(1), 19–32.
Smith, P. (2000). Progressive revelation. In P. Smith (Ed.), *A concise encyclopedia of the Bahá'í faith* (pp. 276–277). Oneworld Publications.
Taheri-Najafabadi, A. (2013). CO2 chemical conversion to useful products: An engineering insight to the latest advances toward sustainability. *International Journal of Energy Research, 37*(6), 485–499.
Taylor, B., Van Wieren, G., & Zaleha, B. (2016). The greening of religion hypothesis (part two): Assessing the data from Lynn White, Jr, to pope Francis. *Journal for the Study of Religion. Nature and Culture, 10*(3), 306–378.
Tomalin, E., Haustein, J., & Kidy, S. (2019). Religion and the sustainable development goals. *The Review of Faith & International Affairs, 17*(2), 102–118.
Tu, C., Suweis, S., & D'Odorico, P. (2019). Impact of globalization on the resilience and sustainability of natural resources. *Nature Sustainability, 2*(4), 283–289.
White, R. (1989). Spiritual foundations for an ecologically sustainable society. *The Journal of Bahá'í Studies, 2*(1), 33–57.
Zagonari, F. (2021). Religious and secular ethics offer complementary strategies to achieve environmental sustainability. *Humanities and Social Sciences Communications, 8*(1), 1–13.

Index

N. Singh et al. (eds.), *Faith Traditions and Sustainability*, Management, Change, Strategy and Positive Leadership, https://doi.org/10.1007/978-3-031-41245-5

MIX
Papier aus verantwortungsvollen Quellen
Paper from responsible sources
FSC® C105338

If you have any concerns about our products,
you can contact us on
ProductSafety@springernature.com

In case Publisher is established outside the EU,
the EU authorized representative is:
Springer Nature Customer Service Center GmbH
Europaplatz 3, 69115 Heidelberg, Germany

Printed by Libri Plureos GmbH
in Hamburg, Germany